HUMAN-IN-THE-LOOP LEARNING AND CONTROL FOR ROBOT TELEOPERATION

HUMAN-IN-THE-LOOP LEARNING AND CONTROL FOR ROBOT TELEOPERATION

CHENGUANG YANG
Bristol Robotics Laboratory
University of the West of England
Bristol, United Kingdom

JING LUO
Wuhan University of Technology
Wuhan, China

NING WANG
Bristol Robotics Laboratory
University of the West of England
Bristol, United Kingdom

ACADEMIC PRESS

An imprint of Elsevier

ELSEVIER

Academic Press is an imprint of Elsevier
125 London Wall, London EC2Y 5AS, United Kingdom
525 B Street, Suite 1650, San Diego, CA 92101, United States
50 Hampshire Street, 5th Floor, Cambridge, MA 02139, United States
The Boulevard, Langford Lane, Kidlington, Oxford OX5 1GB, United Kingdom

Notices

Knowledge and best practice in this field are constantly changing. As new research and experience
broaden our understanding, changes in research methods, professional practices, or medical
treatment may become necessary.

Practitioners and researchers must always rely on their own experience and knowledge in
evaluating and using any information, methods, compounds, or experiments described herein. In
using such information or methods they should be mindful of their own safety and the safety of
others, including parties for whom they have a professional responsibility.

To the fullest extent of the law, neither the Publisher nor the authors, contributors, or editors,
assume any liability for any injury and/or damage to persons or property as a matter of products
liability, negligence or otherwise, or from any use or operation of any methods, products,
instructions, or ideas contained in the material herein.

ISBN: 978-0-323-95143-2

For information on all Academic Press publications
visit our website at https://www.elsevier.com/books-and-journals

Publisher: Mara E. Conner
Acquisitions Editor: Sonnini R. Yura
Editorial Project Manager: Fernanda A. Oliveira
Production Project Manager: Kamesh R
Cover Designer: Mark Rogers

Typeset by VTeX

Working together
to grow libraries in
developing countries

www.elsevier.com • www.bookaid.org

Contents

Author biographies

Prof. Chenguang Yang received the PhD degree in control engineering from the National University of Singapore, Singapore, in 2010, and post-doctoral training in human robotics from the Imperial College London, London, UK. He was awarded UK EPSRC UKRI Innovation Fellowship and individual EU Marie Curie International Incoming Fellowship. As the lead author, he won the IEEE Transactions on Robotics Best Paper Award (2012) and IEEE Transactions on Neural Networks and Learning Systems Outstanding Paper Award (2022). He is the Corresponding Co-Chair of IEEE Technical Committee on Collaborative Automation for Flexible Manufacturing, a Fellow of Institution of Mechanical Engineers (IMechE), a Fellow of Institute of Engineering and Technology (IET), and a Fellow of British Computer Society (BCS). His research interest lies in learning and control of mechatronic and robotic systems.

Dr. Jing Luo received his PhD degree in robotics and control from South China University of Technology and acted as a researcher at the LIRMM, Montpellier, France, and the Imperial College of Science, Technology and Medicine, London, UK. He is involved in several scientific research projects and has produced several research results that have been published in several automation and robots related journals. His research interests include robotics, teleoperation, wearable devices, and human–robot interaction.

Dr. Ning Wang received her MPhil and PhD degrees in electronics engineering from the Chinese University of Hong Kong. She has rich project experience and has been a key member of EU, UK EPSRC, and industrial projects. She has won several awards, including the IET premium award for best paper 2022, the best paper award of ICIRA'15 and the award of merit of the 2008 IEEE Signal Processing Postgraduate Forum, and she was nominated for the ISCSLP'10 best student paper award. She is currently an associated editor of the *International Journal of Advanced Robotic Systems*, and *Frontiers in Computer Science*, and she served as invited speaker at ROMAN'20 and EECR'23. Her research interests lie in human–robot interaction, machine learning, and autonomous driving.

Preface

Robots are utilized more and more frequently nowadays. They are employed in industrial production processes and human daily life with many more promising prospects for applications. As the depth and breadth of robotics applications continue to expand, robots are expected to work in dynamic and unknown environments to meet more demanding requirements for complex and diverse tasks. Remote control of robots with semi-autonomous capabilities, also known as telerobotics, is a popular and well-developed area which has attracted considerable attention from academia and industry. Telerobots are widely used for missions in hazardous environments, because state-of-the-art autonomous robots cannot perform all required by themselves. Many relevant algorithms and techniques have been developed to address the limitations of physical human–robot interaction (pHRI), perception, and learning capabilities in telerobot development. Among these approaches, teleoperation and robot learning have been proven to enhance performance of telerobots with higher stability and efficiency. However, most of these methods focus only on the studies of the characteristics of the telerobots themselves, without taking into account the dominant role and intelligence of human operators in the closed-loop system.

In this book, we will present our recent work on control and learning methods, taking into account the human factors to enable telerobots interacting with humans in a user-friendly and intelligent way. Different aspects of teleoperation control will be investigated in terms of uncertainty compensation, user experience, shared control, and pHRI. Learning plays an important role to help telerobots acquire human manipulation skills, especially in human-in-the-loop teleoperating systems, allowing humans to focus mainly on high-level cognitive tasks such as decision-making.

This book will focus primarily on learning and control technologies specifically developed for human-in-the-loop systems that can improve telerobots' performance. The organization of the book is summarized below.

Chapter 1 will provide an overview of typical remote operating systems. Teleoperating systems include unilateral teleoperation, bilateral teleoperation, and multilateral teleoperation, which can help human operators to accomplish tasks in complex situations. Given the importance of pHRI in teleoperation, we will introduce both unimodal and multimodal inter-faces of pHRI that are used to provide human operators with perceptual feedback for more natural and effective interactions with robots. We will also present learning and control algorithms, which will be used to obtain human-like compliant operations. Next, we will present typical examples of telerobot applications such as the da Vinci surgical robot, the Canadian robot, the Kontur-2 project, and Robonaut.

In Chapter 2, we will present the platforms and software systems used for teleoperation in this book. The teleoperation platforms include mobile robots, Baxter robots, KUKA LBR iiwa robots, haptic devices, and different sensors such as Kinect, Mini45 force/torque sensors, and the MYO Armband. The software systems for teleoperation are used to describe and analyze kinematic and dynamic models of robots and various simulation applications. The main robotics software systems include the OpenHaptics Toolkit, MATLAB® Robotics Toolbox, the Robotics Operating System, Gazebo, and Coppeliasim.

Chapter 3 will focus on how to control self-driving robots in the presence of uncertainty. Wave variables and neural learning methods show great benefits in dealing with time delays, especially in uncertain systems and environments. In this chapter, basic teleoperation control based on bilateral teleoperation systems, such as position-to-position control and four-channel control, will be introduced. Neural learning control methods will be presented to improve teleoperation performance in the presence of nonlinearity and uncertainty. In addition, wave-variable correction-based methods and multi-channel-based wave-variable methods will be intro-duced due to their construction and passivity. Several experimental case studies show better performance in comparison with traditional control methods, such as traditional PD control with/without wave variable methods.

In Chapter 4, a user experience-based teleoperation control scheme is proposed to improve the reliability and controllability of the system. Control algorithms such as electromyography (EMG)-based methods are applied to adjust the control gain of the telerobot for personalized interaction with the environment. Variable control with tremor attenuation is utilized to ensure the stability and accuracy of the teleoperation system. Our work combines variable impedance control and virtual fixtures to guide human operation. In addition, variable gain control based on the Lyapunov–Krasovskii functions is developed to approximate the effects caused by kinetic uncertainties and disturbances. Virtual reality (VR)-based teleoperation applications and experimental case studies are presented to demonstrate the effectiveness of the user experience approaches.

Chapter 5 describes the shared control algorithm, which combines autonomous robot control and human–robot remote control. Collision avoidance control allows the robot to avoid any potential collisions or known obstacles while working, thus ensuring the safety of both humans and the robot. Electroencephalography (EEG)–based visual fusion and thought control make it easier for users to operate the robot's manipulators. Mixed reality (MR)-based shared control is proposed for path planning of omnidirectional mobile robots. EMG-based fragmentation control methods with target-free obstacle avoidance and artificial potential fields are used to solve the obstacle avoidance problem and motion control. The performance of these methods is demonstrated experimentally.

In Chapter 6, HRI-based control is applied to improve the perceptiveness and intelligence of the remote robot to reduce the operator's operational burden and stress. By considering the interaction situation between robot and human, such as trajectory and interaction force, the autoregression-based model can accurately predict the human motion intention. A linear guided virtual fixture (VF) based on actual motion characteristics and EMG is introduced to provide force feedback to the human operator, which in turn helps the operator to perform the task more accurately and autonomously. This chapter concludes with human motion prediction and EMG-based VF experiments to show the effectiveness of the proposed method.

Chapter 7 will propose a task learning scheme for a human-in-the-loop teleoperation system for improving the efficiency of robot learning and the automatic generation of tasks. For task learning, the space vector approach and dynamic time warping method are presented to process the demonstration data in 3D space. Then, several robot task trajectory learning methods based on machine learning are proposed, such as the Gaussian mixture model and Gaussian mixture regression, the extreme learning machine, locally weighted regression, and the hidden semi-Markov model, which do not require human involvement after learning human skills, to improve the efficiency. Several experiments are executed successfully to show the performance of our methods.

CHAPTER ONE

Introduction

1.1. Overview of typical teleoperation systems

1.1.1 What is teleoperation in robotics?

Teleoperation, also called telerobotics, is a technical term for the remote control of robots. The prefix "tele-" in teleoperation means "long distance," which means that the human operator can control the robot's actions from remote sides.

A teleoperation system can execute complex tasks in dangerous situations. For example, in the nuclear industry, human workers cannot work for a long time in radioactive environment. The teleoperation systems allow workers to control remote robots in an effective way.

Furthermore, since the COVID-19 pandemic began, the use of teleoperation systems has grown significantly, primarily due to the rising demand for contactless delivery and medical treatment. In these situations, face-to-face communication and interaction are not avoided to ensure safety.

1.1.2 Composition of a typical teleoperation system

A typical teleoperation system is commonly composed of five interconnected elements: at least one leader robot that is operated by the human, at least one follower robot that performs the operations in the environment, a remote workspace, a human operator, and communication channels. A block diagram of a typical teleoperation system is illustrated in Fig. 1.1.

Unilateral teleoperation

In unilateral teleoperation, one leader robot and one follower robot are provided, and the information flows in one direction. The leader site sends only the necessary commands (like positions and velocities) to the follower site through the communication lines, and these commands are performed in the control system of the follower site [1].

Perhaps this type of teleoperation system is not significantly different from the ordinary control application (where the leader site only transmits user commands). Applications of unilateral teleoperation seem to be scarce.

1

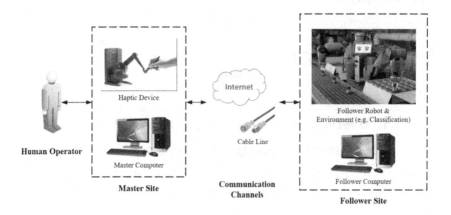

Figure 1.1 Block diagram of a typical teleoperation system.

Bilateral teleoperation

In bilateral teleoperation, the follower robot possesses force sensors, thus the reaction force can be transmitted from the task site back to the leader site. Although transmitting the force back to the human operator enables the human to rely on their haptic sense, it may cause instability in the system due to time delay in the communication lines. Bilateral teleoperation is more challenging than unilateral teleoperation. Traditional and potential applications range from space and undersea exploration to telesurgery [2].

Multilateral teleoperation

As the field of teleoperation grows, multilateral frameworks have received much attention during the past years. A multilateral teleoperation framework contains more than two robotic agents (leader and follower robots), allowing collaborative scenarios between the leader site and the follower site as with human behavior [3].

1.2. Human–robot interaction in teleoperation

1.2.1 Why emphasize human–robot interaction?

In typical teleoperation systems, the human operator can only perceive the environment by simple position/force feedback, which leads to a lack of perceptive information like visual, haptic, and auditory information for the human operator to understand the surrounding environment and make optimal decisions. For example, in in-contact tasks, the operator may exert a large force on a breakable item without haptic information. Thus, a perception system that provides this perceptive feedback to the human operator is needed in modern teleoperation.

1.2.2 Several unimodal feedback for better interaction

Virtual reality

In virtual reality (VR), the real world can be replaced by a computer-generated world, which can give the human operator a sense of immersion. VR systems are based on input devices to interact with the operator to receive commands and capture data from the real world and output devices to display the responses in the virtual world during the interaction; for example, headmounted displays (HMDs) present this information in a visual way (adapted from Ref. [4]).

In Ref. [5], the authors developed a VR-based robot control system with manipulation assistance, which allows the human operator to specify high-level manipulation goals effectively.

Augmented reality

Augmented reality (AR) is a form of human–computer interaction (HCI) that superimposes information created by computers over a real environment [4]. Instead of replacing the real scene by a virtual one as in VR, AR enriches the surrounding environment with processed information, which can be applied to any of the human senses (usually in a visual way) and expand perception.

In teleoperation environments, AR has been used to complement human sensorial perception in order to help the operator perform teleoperation tasks [4]. For example, in the applications of AR in teleoperation tasks, virtual tools used for guidance were provided in Ref. [6].

Haptic feedback

While processing in-contact teleoperation tasks which require force control, haptic feedback is essential. Haptic feedback has been well-studied in the teleoperation field. In Ref. [7], the authors proposed a robotic teleoperation system with wearable haptic feedback for telemanipulation in cluttered environments. Moreover, haptic interfaces are also employed in precise telemanipulation, such as surgical robots, micromanipulation, and micro-assembly.

EMG

The electromyography (EMG) signal is a biomedical signal that measures electrical currents generated in muscles during contraction. Since the movement and force of the human body are inseparable from muscle contraction, EMG signals can be used to estimate both movement and force exerted by the operator.

In the teleoperation system, the utilization of EMG signals is an effective way to enhance the operation experience, since traditional wearable devices are replaced with more natural and contactless sensors. In Ref. [8], EMG signals from the human arm were used to control a robot manipulator, since muscles are responsible not only for moving the human limbs but also for exerting forces on the environment.

1.2.3 Multimodal interface for better teleoperation

The multimodal interfaces in teleoperation aim to provide immersive solutions and enhance the overall human performance. Extensive comparisons have been made in Ref. [9] to show that regardless of task complexity, multimodal interfaces could improve performance. Research in cognitive psychology suggests that multisensory stimuli enhance human perceptual learning. The structure of a multimodal teleoperation system is illustrated in Fig. 1.2.

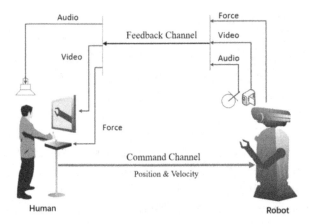

Figure 1.2 Structure of a multimodal teleoperation system adapted from Ref. [10].

1.3. Learning and control for teleoperation

1.3.1 Introduction of LfD

Nowadays, robot manipulators are widely used to perform tasks in certain structured environments due to the advantages of low cost, high efficiency, and safety. However, it is still difficult to introduce them into physical in-contact tasks, since these tasks require robots not only to track the desired trajectories, but also to interact with the environment physically. Thus, manipulators with human-like and complete manipulation skills are expected.

Learning from demonstration (LfD) is an effective way to transfer skills from human to robots, which can be understood as learning skills by observing the demonstration by human. According to the demonstration approach in Ref. [11], LfD can be divided into three categories based on demonstration: kinesthetic teaching, teleoperation, and passive observation, as shown in Fig. 1.3.

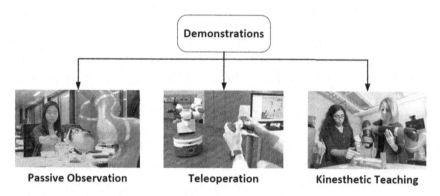

Figure 1.3 Categorization of LfD methods based on the demonstration. Adapted from Ref. [11].

In kinesthetic teaching and passive observation, the human operator must cooperate with a robot manipulator, and smooth trajectories are required to achieve good effects. These requirements limit the use of LfD in complex and dangerous working environments. Learning from tele-operation can solve the aforementioned issues effectively. Furthermore, immersive teleoperation enables the human operator to teach robots with more natural and realistic demonstrations.

Skill learning based on trajectory matching is a hotspot in LfD research. Typical skill learning approaches in LfD include the dynamical system approach and the probability and statistical approach.

1.3.2 Skill learning in teleoperation

Dynamical system approach

In the dynamical system approach, a trajectory is generated by the evolution of the dynamic system over time and space, and demonstration data are used for training dynamic systems. In Ref. [12], a stable estimator of dynamical systems (SEDS) based on Gaussian mixture models (GMMs) was proposed to learn the parameters of the dynamic system, and it is employed to model various motions, such as playing minigolf in Ref. [13] and human handwriting motion in Ref. [12]

Besides, the dynamic movement primitive (DMP) is another framework to realize movement planning, online trajectory modification for LfD use, which was originally proposed by Ijspeert et al. [14,15]. By using the superposition of a spring–damper system and a nonlinear function, DMP can encode the demonstration trajectory in an effective way. Compared with traditional means of encoding trajectories, such as spline-based decomposition, the DMP encoding skills have a variety of benefits [16]. For example, DMPs can guarantee the system's global stability, and the velocity of motion can be adapted by changing the time constant parameter.

However, the original DMP also has limitations with respect to motion planning in some situations, for example, the goal point coinciding with the start point in any dimension. The original DMP cannot achieve the force control of robots for contact tasks, such as assembly. Therefore, since the original DMP is proposed, a variety of modified DMPs have been proposed to tackle these limitations, like DMPs with perceptual term (DMPP), coupling DMPs, and reinforcement learning (RL)-based DMP. In Ref. [17], the author extended the DMP framework with sensory feedback for movement adaptation in grasping tasks. In Ref. [18,19], the authors took repulsive potential fields as coupling terms into DMPs for obstacle avoidance. In Ref. [20], deep RL and a hierarchical strategy were used to optimize and generalize the skills produced by DMPs.

Probability and statistical approach

In the probability and statistical approach, the motion is modeled as a random model, such as GMM, hidden Markov model (HMM), or dynamic Bayesian network (DBN), to realize trajectory modeling and trajectory matching. Since the random model can deal with noise, LfD based on the probability and statistical approach has a strong noise processing ability.

GMM has been employed to model the joint distribution of input variables and demonstrated trajectories [21]. In Ref. [22], GMM was used to model both movement and force patterns for robots learning impedance behaviors. Usually, GMM is complemented with Gaussian mixture regression (GMR) to retrieve the desired trajectory. As an extension of GMM, a task-parameterized formulation is studied in Ref. [23], which in essence models local (or relative) trajectories and corresponding local patterns, therefore endowing GMM with better extrapolation performance. In Ref. [24], GMM was utilized to encode and parameterize the smooth task trajectory to realize a task learning mechanism of the telerobots.

1.3.3 Control issues in teleoperation

Stability and transparency

There are two main objectives in control issues for a teleoperation system: stability and transparency. Obviously, stability is the fundamental requirement of the system, despite the existence of time delay in communication channels and signal noise. Transparency means that the interaction between the human operator and the remote environment is natural and comfortable, which implies a feeling of presence at the remote site [4]. Ideal transparency means that the human operator can operate the follower robot and feel to be personally at the remote scene.

Since the unilateral teleoperation cannot provide information feedback for operators, the telepresence and stability discussed here will be based on bilateral teleoperation robots.

Bilateral teleoperation structure

Based on the concept of circuit, the bilateral system can be equivalent to a dual-port network. The force and movement signals sent from the human operator to the leader robot can be regarded as the input voltage and current, and the force and movement signals of the follower robot in the environment can be regarded as the output voltage and current. The network includes the controller at both sites and the communication channels.

The dual-port network above can be transformed to a 2-by-2 matrix as follows, which can be used for stability verification:

$$\begin{bmatrix} F_h \\ -X_e \end{bmatrix} = \begin{bmatrix} h11 & h12 \\ h21 & h22 \end{bmatrix} \begin{bmatrix} X_h \\ F_e \end{bmatrix}, \tag{1.1}$$

where F_h is the interaction force between the operator and the leader robot, F_e is the interaction force between the follower robot and the environment, and X_h and X_e are the movement of the leader robot and the follower robot.

Stability guarantee

In bilateral teleoperation, the Raisbeck passivity theorem [25] and the Llewellyn absolute stability theorem [26] are two famous theorems to analyze stability. In these two theorems, the mathematical models of the human operator and the environment do not need to be clear, but it must be ensured that they are passive.

Control strategy: shared control

According to the control mode, teleoperation systems can be divided into three categories: direct control, supervised control, and shared control [27]. In direct control, the follower robot is directly controlled by the human operator without autonomous abilities. In supervised control, the robot executes the tasks according to the pre-programmed code, in which the human merely supervises the execution process.

Shared control is a hybrid strategy that combines direct control and supervised manipulation, in which the human operator collaborates with robots based on a mechanism. The shared control framework has been well studied, for example, human–robot shared manipulation in Ref. [28] and telecontrol of the mobile robot's motion and obstacle avoidance in Ref. [29]. Similarly, in Ref. [30], a human–robot shared control strategy was developed to achieve autonomous obstacle avoidance in manipulation.

Control strategy: compliance control

When the robot performs an in-contact task in uncertain dynamics exist, like assembly, polishing, and robot-assisted echography, stiffness and damping of the robot are required to be changed to achieve the robot's behavior during the interaction.

Impedance control is an important control architecture for compliant control. By using the real motion of the robot as input and the interaction force as output, impedance control can maintain an impedance relationship between interaction force and motion. Besides, the impedance can even be adjusted based on various tasks. The variable impedance control is well studied to deal with the in-contact task under less predictive and structured environments. Hogan initially studied impedance control for manipulators [31], and since then, a number of improved methods were proposed to deal with various challenges in robotic control. In addition, Yang et al. proposed a human-like learning controller to achieve variable impedance when robots interact with unknown environments [32]. In Ref. [33], the authors studied the stability considerations for variable impedance control.

1.4. Project cases of teleoperation

The technique of teleoperation is mainly used in the surgical field, the aerospace field, etc. Several famous cases which apply teleoperation are introduced below.

1.4.1 The Da Vinci surgical robot

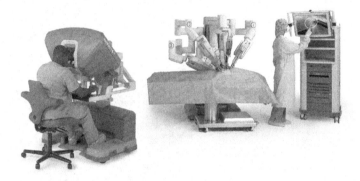

Figure 1.4 The Da Vinci teleoperated robotic system [34].

The Da Vinci surgical system illustrated in Fig. 1.4 is a robotic surgical system made by the American company Intuitive Surgical. It is designed to facilitate surgery using a minimally invasive approach and is controlled by a surgeon from a console. The robot is mainly used for prostatectomies and increasingly for cardiac valve repair, etc. [34].

The Da Vinci teleoperated robotic system is based on a leader–follower control concept and it has two units. One of the units is the surgeon's console unit, which consists of the display system, the handle shown in Fig. 1.5(b) serves as the surgeon's user interface, and the controller. The other unit is the patient side cart composed of four follower manipulators, three for telemanipulation of surgical instruments and one dedicated to the camera, as shown in Fig. 1.5(a).

The Da Vinci surgical system provides a human–robot interface that directly connects the movement of the surgeon's hand to the movement of the manipulator's end-effectors. The surgeon can visualize stereoscopic images via the 3D display above the hands and the vision cart. The surgeon can use the Da Vinci handle to remotely move the instrument's tip. The surgeon controls the follower sitting on a stool by the console, which is positioned remotely from the patient. The console serves as an interface between the surgeon and the Da Vinci robot and the surgeon views the operation within the console hood [35]. If a surgeon leaves the console, the Da Vinci robot will automatically stop moving. Furthermore, the controller can filter out surgeon tremor, making the instrument tips steadier and surgical operations safer compared to traditional laparoscopic instruments.

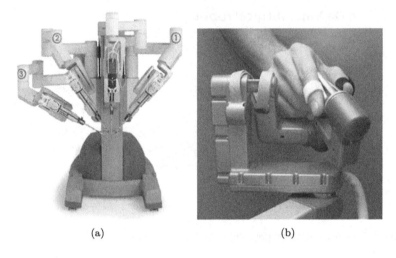

(a)   (b)

Figure 1.5 The patient side cart (a) and the Da Vinci handle (b) [34].

1.4.2 Canadian Robot

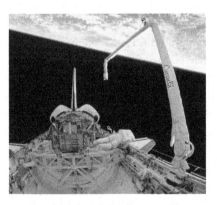

Figure 1.6 Canadian Robot (right) during a space shuttle mission [36].

The Canadian Robot illustrated in Fig. 1.6 is the remote-controlled mechanical arm provided by Canada to NASA for installation on the shuttle

Orbiter. During its career with NASA's Space Shuttle System, the robotic arm performed tasks such as deploying, capturing and repairing satellites, as well as positioning astronauts, maintaining equipment, and moving cargo.

The Canadian Robot is 15.2 m long and has six joints that correspond to the joints of the human arm. The end-effector is the unit at the end of the wrist that grapples the payload's grapple fixture.

The remote manipulator system enables the operator to command motion of the tip of the arm without having to give conscious thought to the motions of the six individual joints. This motion may be commanded by his use of the translation and rotation rate command hand controllers. The TV cameras provide visual cues to the astronaut or the operator controlling the robotic arm from inside the shuttle. There are hand controllers, display panels, and a signal-processing interface box in the control station of the shuttle [36].

With a bigger and more intelligent version than the original, Canadian Robot2 played a more important role in the construction of the International Space Station. It can work together with a special-purpose dexterous manipulator called Dextre which can be moved around at the end of Canadian Robot2.

1.4.3 The Kontur-2 Project

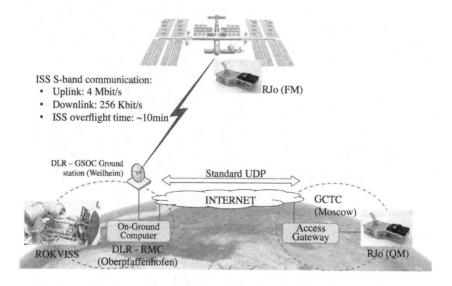

Figure 1.7 Kontur-2 scenarios [37].

The Kontur-2 Project is a project for the in-flight verification of force feedback and telepresence technologies, and the space scenarios of this project are shown in Fig. 1.7. Its main objectives are the development of a space-qualified two-degrees of freedom force feedback joystick [37], the implementation of telepresence technologies, and the investigation of the overall performance when the teleoperating robotic systems on Earth from space with force feedback.

In order to achieve the next milestone in planetary exploration missions to allow astronauts to work with robots on the ground from the orbital station, a new technology for real-time telemanipulation is used in Ref. [38].

Furthermore, the force feedback joystick was installed in the Russian module of the International Space Station (ISS) in August 2015 and operated until December 2016. In August 2015 the first experiments were conducted successfully.

1.4.4 Robonaut

Robonaut is a NASA robot which is designed to be humanoid. The core idea behind it is to have a humanoid machine work alongside astronauts. Its form factor and dexterity are designed such that Robonaut can use space tools and work in similar environments suited to astronauts [39] (Fig. 1.8(a)).

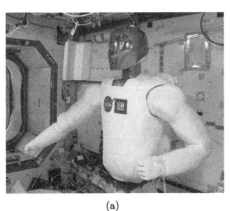

(a)   (b)

Figure 1.8 Robonaut2 working in the space station [40].

The latest Robonaut version is R2, which was delivered to ISS in 2011. R2 could help assemble and service space science satellites in orbit beyond the Space Shuttle's reach, and it could handle long-duration exposed payloads on the space shuttle or ISS. Furthermore, it could complement the work of larger robots or serve as an astronaut's assistant during the space walks (Fig. 1.8(b)).

Robonaut can function in two ways. One is accomplishing simple tasks by itself. The other is operating by remote control, which is also called teleoperation [40]. Robonaut provides a full-immersion environment for astronauts, who can wear the VR-based telepresence gloves and helmet, so that the operator's hand, arm, and neck can be directly connected to the Robonaut system. The helmet enables the astronaut to see through the robot's stereo camera head. The sensors in the gloves determine finger positions and send commands to the Robonaut hand. The Robonaut system will integrate new force and tactile feedback devices to give the operator natural cues to the system's force amplitude and direction [41].

1.5. Conclusion

In this chapter, we gave an overview of the typical teleoperation systems, including the meaning of teleoperation and the composition of a typical teleoperation system. Then a number of unimodal and multimodal interfaces are introduced, which aim to provide a better human–robot interaction. Furthermore, human–robot skill transfer (i.e., the dynamical system approach and the probability and statistical approach) and control algorithms (i.e., shared control and impedance control) in teleoperation are introduced. Finally, four successful and famous projects of teleoperation are analyzed, including the Da Vinci surgical robot, the Canadian Robot, the Kontur-2 project, and Robonaut.

References

[1] Mehmet Dede, Sabri Tosunoglu, Fault-tolerant teleoperation systems design, Industrial Robot: An International Journal (2006).
[2] Peter F. Hokayem, Mark W. Spong, Bilateral teleoperation: An historical survey, Automatica 42 (12) (2006) 2035–2057.
[3] Mahya Shahbazi, Seyed Farokh Atashzar, Rajni V. Patel, A systematic review of multilateral teleoperation systems, IEEE Transactions on Haptics 11 (3) (2018) 338–356.
[4] Luis Basañez, Raúl Suárez, Teleoperation, Springer Berlin Heidelberg, Berlin, Heidelberg, 2009, pp. 449–468.
[5] Christian Barentine, Andrew McNay, Ryan Pfaffenbichler, Addyson Smith, Eric Rosen, Elizabeth Phillips, A VR teleoperation suite with manipulation assist, in: Companion of the 2021 ACM/IEEE International Conference on Human–Robot Interaction, 2021, pp. 442–446.
[6] Henry Portilla, Luis Basanez, Augmented reality tools for enhanced robotics teleoperation systems, in: 2007 3DTV Conference, IEEE, 2007, pp. 1–4.

[7] Joao Bimbo, Claudio Pacchierotti, Marco Aggravi, Nikos Tsagarakis, Domenico Prattichizzo, Teleoperation in cluttered environments using wearable haptic feedback, in: 2017 IEEE/RSJ International Conference on Intelligent Robots and Systems (IROS), IEEE, 2017, pp. 3401–3408.

[8] Panagiotis K. Artemiadis, Kostas J. Kyriakopoulos, EMG-based position and force control of a robot arm: Application to teleoperation and orthosis, in: 2007 IEEE/ASME International Conference on Advanced Intelligent Mechatronics, IEEE, 2007, pp. 1–6.

[9] Eleftherios Triantafyllidis, Christopher Mcgreavy, Jiacheng Gu, Zhibin Li, Study of multimodal interfaces and the improvements on teleoperation, IEEE Access 8 (2020) 78213–78227.

[10] Xiao Xu, Michael Panzirsch, Qian Liu, Eckehard Steinbach, Integrating haptic data reduction with energy reflection-based passivity control for time-delayed teleoperation, in: 2020 IEEE Haptics Symposium (HAPTICS), IEEE, 2020, pp. 109–114.

[11] Harish Ravichandar, Athanasios S. Polydoros, Sonia Chernova, Aude Billard, Recent advances in robot learning from demonstration, Annual Review of Control, Robotics, and Autonomous Systems 3 (2020) 297–330.

[12] S. Mohammad Khansari-Zadeh, Aude Billard, Learning stable nonlinear dynamical systems with Gaussian mixture models, IEEE Transactions on Robotics 27 (5) (2011) 943–957.

[13] Seyed Mohammad Khansari-Zadeh, Klas Kronander, Aude Billard, Learning to play minigolf: A dynamical system-based approach, Advanced Robotics 26 (17) (2012) 1967–1993.

[14] Auke Jan Ijspeert, Jun Nakanishi, Stefan Schaal, Trajectory formation for imitation with nonlinear dynamical systems, in: Proceedings 2001 IEEE/RSJ International Conference on Intelligent Robots and Systems. Expanding the Societal Role of Robotics in the Next Millennium (Cat. No. 01CH37180), vol. 2, IEEE, 2001, pp. 752–757.

[15] Auke Jan Ijspeert, Jun Nakanishi, Stefan Schaal, Movement imitation with nonlinear dynamical systems in humanoid robots, in: Proceedings 2002 IEEE International Conference on Robotics and Automation (Cat. No. 02CH37292), vol. 2, IEEE, 2002, pp. 1398–1403.

[16] Auke Jan Ijspeert, Jun Nakanishi, Heiko Hoffmann, Peter Pastor, Stefan Schaal, Dynamical movement primitives: Learning attractor models for motor behaviors, Neural Computation 25 (2) (2013) 328–373.

[17] Peter Pastor, Ludovic Righetti, Mrinal Kalakrishnan, Stefan Schaal, Online movement adaptation based on previous sensor experiences, in: 2011 IEEE/RSJ International Conference on Intelligent Robots and Systems, IEEE, 2011, pp. 365–371.

[18] Èric Pairet, Paola Ardón, Michael Mistry, Yvan Petillot, Learning generalizable coupling terms for obstacle avoidance via low-dimensional geometric descriptors, IEEE Robotics and Automation Letters 4 (4) (2019) 3979–3986.

[19] Akshara Rai, Franziska Meier, Auke Ijspeert, Stefan Schaal, Learning coupling terms for obstacle avoidance, in: 2014 IEEE-RAS International Conference on Humanoid Robots, IEEE, 2014, pp. 512–518.

[20] Wonchul Kim, Chungkeun Lee, H. Jin Kim, Learning and generalization of dynamic movement primitives by hierarchical deep reinforcement learning from demonstration, in: 2018 IEEE/RSJ International Conference on Intelligent Robots and Systems (IROS), IEEE, 2018, pp. 3117–3123.

[21] Sylvain Calinon, Florent Guenter, Aude Billard, On learning, representing, and generalizing a task in a humanoid robot, IEEE Transactions on Systems, Man and Cybernetics. Part B. Cybernetics 37 (2) (2007) 286–298.

[22] Leonel Rozo, Sylvain Calinon, Darwin Caldwell, Pablo Jiménez, Carme Torras, Learning collaborative impedance-based robot behaviors, in: Proceedings of the AAAI Conference on Artificial Intelligence, vol. 27, 2013.

[23] Sylvain Calinon, A tutorial on task-parameterized movement learning and retrieval, Intelligent Service Robotics 9 (1) (2016) 1–29.

[24] Jing Luo, Chenguang Yang, Qiang Li, Min Wang, A task learning mechanism for the telerobots, International Journal of Humanoid Robotics 16 (02) (2019) 1950009.

[25] Gordon Raisbeck, A definition of passive linear networks in terms of time and energy, Journal of Applied Physics 25 (12) (1954) 1510–1514.

[26] F.B. Llewellyn, Some fundamental properties of transmission systems, Proceedings of the IRE 40 (3) (1952) 271–283.

[27] Jing Luo, Wei He, Chenguang Yang, Combined perception, control, and learning for teleoperation: Key technologies, applications, and challenges, Cognitive Computation and Systems 2 (2) (2020) 33–43.

[28] Reuben M. Aronson, Thiago Santini, Thomas C. Kübler, Enkelejda Kasneci, Siddhartha Srinivasa, Henny Admoni, Eye-hand behavior in human–robot shared manipulation, in: Proceedings of the 2018 ACM/IEEE International Conference on Human–Robot Interaction, 2018, pp. 4–13.

[29] Yanbin Xu, Chenguang Yang, Xiaofeng Liu, Zhijun Li, A teleoperated shared control scheme for mobile robot based SEMG, in: 2018 3rd International Conference on Advanced Robotics and Mechatronics (ICARM), IEEE, 2018, pp. 288–293.

[30] Xinyu Wang, Chenguang Yang, Hongbin Ma, Long Cheng, Shared control for teleoperation enhanced by autonomous obstacle avoidance of robot manipulator, in: 2015 IEEE/RSJ International Conference on Intelligent Robots and Systems (IROS), 2015, pp. 4575–4580.

[31] Neville Hogan, Impedance control: An approach to manipulation: Part I—theory, Journal of Dynamic Systems, Measurement, and Control 107 (1) (1985) 1–7.

[32] Chenguang Yang, Gowrishankar Ganesh, Sami Haddadin, Sven Parusel, Alin Albu-Schaeffer, Etienne Burdet, Human-like adaptation of force and impedance in stable and unstable interactions, IEEE Transactions on Robotics 27 (5) (2011) 918–930.

[33] Klas Kronander, Aude Billard, Stability considerations for variable impedance control, IEEE Transactions on Robotics 32 (5) (2016) 1298–1305.

[34] V. Ferrari, C. Freschi, F. Melfi, Technical review of the Da Vinci surgical telemanipulator, The International Journal of Medical Robotics and Computer Assisted Surgery 9 (4) (2013) 396–406.

[35] Jia Hai Chen, You Li, Jian Ping Gong, Wu Yakun, Application of Da Vinci surgical robotic system in hepatobiliary surgery, International Journal of Surgery and Medicine 4 (2018) 22–27.

[36] Bruce A. Aikenhead, Canadarm and the space shuttle, Journal of Vacuum Science & Technology. A. Vacuum, Surfaces, and Films 1 (2) (1983) 126.

[37] Jordi Artigas, Ribin Balachandran, Cornelia Riecke, Martin Stelzer, Bernhard Weber, Jee-Hwan Ryu, Alin Albu-Schaeffer, Kontur-2: Force-feedback teleoperation from the international space station, in: 2016 IEEE International Conference on Robotics and Automation (ICRA), IEEE, 2016, pp. 1166–1173.

[38] V. Muliukha, A. Ilyashenko, V. Zaborovsky, A. Novopasheniy, Space experiment "Kontur-2": Applied methods and obtained results, in: 2017 21st Conference of Open Innovations Association (FRUCT), 2017, pp. 244–253.

[39] R.A. Peters, C.L. Campbell, W.J. Bluethmann, E. Huber, Robonaut task learning through teleoperation, in: 2003 IEEE International Conference on Robotics and Automation (Cat. No. 03CH37422), vol. 2, 2003, pp. 2806–2811.

[40] M.A. Diftler, J.S. Mehling, M.E. Abdallah, Robonaut 2 – the first humanoid robot in space, in: 2011 IEEE International Conference on Robotics and Automation, 2011, pp. 2178–2183.

[41] R.O. Ambrose, H. Aldridge, R.S. Askew, Robonaut: NASA's space humanoid, IEEE Intelligent Systems and Their Applications 15 (4) (2000) 57–63.

CHAPTER TWO

Software systems and platforms for teleoperation

2.1. Teleoperation platforms

This section introduced the relevant robot hardware. Robot hardware systems include the mobile robot, the Baxter robot, the KUKA LBR iiwa robot, Touch X, Omega.7, the Kinect camera, Mini45, and the MYO Armband.

2.1.1 Mobile robot

Figure 2.1 Multisensor omnidirectional mobile robot.

Fig. 2.1 shows a typical mobile robot with four omni-directional wheels. It can move in all directions and navigate autonomously. It is like a smart car with four wheels, and it has the same perception and movement capabilities as other robots. It can complete autonomous navigation, and it can also solve other tasks, such as mapping and human–machine tracking. It is intelligent automation equipment.

The mobile robot's movement is achieved by four Mecanum wheels (two on the left and two on the right). The Mecanum wheel consists of

Human-in-the-loop Learning and Control for Robot Teleoperation
https://doi.org/10.1016/B978-0-32-395143-2.00006-1
17

a hub and rollers arranged along the hub circumference at an angle to the wheel. The wheel can rotate around its axis, enabling full-directional movement control of the mobile robot through kinetic analysis. The mobile robot is equipped with a sufficient external interface for users, including USB3.0, HDMI, and RS232 serial ports. The mobile robot also provides external power supply interfaces, taking into account external equipment power supply issues. The mobile robot's master PC is loaded with an Ubuntu-based Robot Operating System (ROS) for data processing and advanced computing. The mobile robot is equipped with lidar, ultrasonic sensors, an Inertial Measurement Unit (IMU), an odometer, and other sensors to measure and estimate the robot's motion. Lidar sensors generate maps through information processing, while odometer and IMU data fusion can be used as a priority location estimator for the next data processing step. Mobile robots are now primarily used with simultaneous localization mapping (SLAM) for location environments. After the establishment of the map, the robot needs to be able to carry out autonomous navigation. How to reasonably estimate its location after path planning is currently one of the issues that researchers are discussing widely [1].

2.1.2 Baxter robot

Figure 2.2 Baxter research robot profile.

The Baxter robot (Fig. 2.2) is an intelligent dual-arm collaborative robot, developed by Rethink Robotics in the United States. It has two arms with seven degrees of freedom and one head with two degrees of freedom. Each arm has seven joints: two shoulders, two elbows, and three wrists. The maximum speed is 2 rad/s for the shoulder joints and elbow joints and 4 rad/s for the wrist joints. The peak torque is 50 Nm for the shoulder joints and elbow joints and 15 Nm for the wrist joints. The Baxter robot uses a Series Elastic Actuator to drive the joints on the robotic arm, each equipped with a dedicated torque sensor for position, speed, and torque control. At the same time, the Baxter robot is equipped with a 360-degree surround sonar and infrared rangefinder to sense the surrounding environment, and these sensors allow users to develop a range of applications.

Baxter provides users with a complete Software Development Kit (SDK), standard ROS functional tables, and an abundant Application Programming Interface (API) interface. The robot is usually programmed using ROS on an Ubuntu operating system through the Baxter SDK. ROS is an open-source operating system for robots that has a rich library to implement interprocess messaging, underlying device control, and other services. Users can write individual code and programs in the Baxter robot to improve robot performance, not only by implementing machine learning and computer vision algorithms, but also by applying various control strategies. Nowadays, many robotics studies have been carried out on Baxter robots, such as neural learning-based telerobot control [2], a robot learning with adaptive neural control [3], and teleoperation control using IMU-based motion capture [4].

The superior performance of the Baxter robot has allowed for new ways of conducting research in robotics. Because robots are equipped with a wealth of intelligent sensing components, researchers can conduct more extensive research on robots, not limited to kinetics and electromechanics, but also machine vision. In addition, the Baxter robot's versatility and strong environmental adaptability also make it possible to carry out production work in a specific environment replacing humans in the future, so it is of practical significance to study it. Today, many well-known research laboratories around the world use Baxter robots as a platform for research, and many promising results have been achieved [5,6].

2.1.3 KUKA LBR iiwa robot

As the very first robot to be mass-produced for physical human–robot collaboration (HRC), the KUKA LBR iiwa robot (Fig. 2.3) offers new solutions for production in the automation industry. LBR iiwa is an industrial robot with the characteristics of high sensitivity, safety, flexibility, and precision. HRC can solve non-ergonomic tasks.

Figure 2.3 KUKA LBR iiwa robot profile.

The machine arm has two models with load capacities of 7 kg and 14 kg. The LBR iiwa has seven axes which are equipped with integral torque sensors that respond to slight external forces and slow down their speed in unexpected contact to reduce kinetic energy damage. The motor is connected to the harmonic drive gear unit with elastic elements on all seven axes, thus ensuring high transmission efficiency and transmission accuracy while also allowing the robot to move more smoothly. The LBR iiwa robot is also equipped with a torque sensor to identify contacts, immediately reducing force and speed, and the robot can use position and buffer control to ensure that sensitive workpieces are handled without injury. Combined with the high-performance servo control, it can follow contours quickly and sensitively.

Researchers can program, start-up, and debug robots through the KUKA Sunrise.Workbench engineering suite. LBR iiwa is programmed in the Java language to maximize the security and productivity of the developers' control of LBR iiwa through KUKA smartPAD [7], which has a variety of visualization features.

2.1.4 Haptic devices

Touch X (Fig. 2.4) is a haptic device made by 3D systems that integrates true 3D navigation and force feedback. Touch X accurately measures the spatial position of 3D space (using the x-, y-, and z-axes) and the spatial orientation of the handheld pen which is mounted on the device (flip up and down, sway left and right, and move sideways). Touch X has high accuracy, and it also has force feedback of three degrees of freedom and six degrees of freedom. Much work has been done with Touch X and Geomagic software.

Figure 2.4 Touch X device.

Touch X has a 160 mm × 120 mm × 120 mm deep workspace, with a range of movements limited to hand movements on the rotating wrist. The device's nominal position resolution is 1100 dpi, which means that when the pen moves about 0.023 mm, the cursor will move one pixel on the screen. Touch X has six degrees of freedom and three axes of force feedback, and when the computer-side cursor interacts with the 3D model of the virtual space, the motor on the device generates power to push the user's hand, thus simulating a touch that allows the user to better perceive the model. Touch X has a six degrees of freedom position sensor that uses a digital encoder to measure the x-axis, the y-axis, and the z-axis in 3D space, and it also has a magnetic absolute position sensor which has 14-bit precision to measure the pen's pitch, roll, and yaw angles for greater accuracy.

The user only needs to use a USB cable to connect Touch X to a computer. People typically program and manipulate haptic devices on Windows systems using OpenHaptics Edition Developer. The Haptics toolkit is patterned after the OpenGL API, so developers can not only use existing OpenGL code to develop geometry, but also the Haptics command to simulate haptic materials with properties such as stiffness and friction. Touch X is widely used in 3D modeling, medicine, gaming, and other fields [8].

Omega.7 (Fig. 2.5) is a haptic force feedback device manufactured by Force Dimension. Omega.7 has a unique active grasping extension. Its end-effector covers the natural range of motion of the human hand and is compatible with a bimanual teleoperation console design. Omega.7 combines full gravity compensation and driftless calibration to improve user comfort and accuracy.

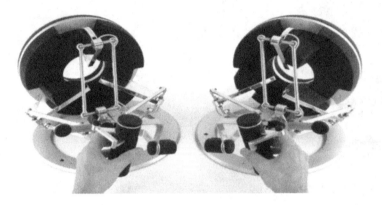

Figure 2.5 Omega.7 device.

The translation of Omega.7's workspace is 160 mm × 110 mm, the rotation is 240 × 140 × 180 degrees, and the grasping is 25 mm. The active gripper can generate a grasping force of up to 8 N in both directions. The interaction mechanism of Omega.7 has seven degrees of freedom and includes a delta parallel mechanism, a wrist joint mechanism, and a clamping hand mechanism. The delta parallel mechanism provides three degrees of freedom for the translational motion to capture the moving position of the operator's hand. The wrist joint mechanism offers three degrees of freedom for rotational motion. The gripper mechanism provides one degree of freedom for the rotational gripping motion for the collection of the user's finger, and the gripper mechanism is arranged on the wrist joint end member. Omega's parallel mechanical design increases the hardness and robustness of the overall structure, making it a rugged and reliable point-contact haptic device. The user only needs to use a USB cable to connect Omega.7 to a computer. Human typically control haptic devices on Windows systems using haptic SDK and robotic SDK. At present, Omega.7 is widely used in the application and research of medical robots, aerospace robots, and microrobots.

2.1.5 Sensors

Kinect (Fig. 2.6) is an RGBD depth camera widely used to obtain deep images, with dynamic capture, image recognition, microphone input, and more features. The advent of the Kinect has made it easier to extract features based on depth maps, so many applications such as virtual reality and video fusion have emerged. At the same time, Kinect is also widely used in the study of robotics.

Figure 2.6 Kinect camera.

Kinect consists of a color camera, a microphone array, and a depth sensor. Compared to Kinect V1 using light coding, the depth sensor of Kinect V2 uses the Time of Flight (TOF) approach to obtain depth information from the returning time of the projected infrared light, which is the biggest difference from a normal camera. These depth images not only have improved resolution, but also have much better depth image quality than the previous generation (Kinect V1).

By using the Kinect 2.0 SDK developed by Microsoft, users can obtain higher-resolution depth images (and color images) using Kinect V2. Kinect V2 has three coordinate spaces: colon space, depth space, and camera space. People can obtain 2D information for a point in a Kinect color image and then combine the depth information in Kinect depth space to obtain 3D camera space. Applications of camera space provided by Kinect V2 include human skeleton tracking and 3D reconstruction [9,10].

Mini45 (Fig. 2.7) is a torque sensor produced by ATI Industrial Automation, which is a monolithic sensor capable of measuring force and torque in six degrees of freedom. Mini45 is widely used in scientific research on remote robots, robotic surgery, and robotic hands.

Figure 2.7 Mini45 force/torque sensor.

Mini45 is available in IP65 and IP68 versions, of which the IP65 version only prevents water splashing, but the IP68 version can be placed in clear water at a depth of 10 m. Mini45 is small and compact, and has a through-hole design that can withstand approximately 44 N of pulling forces through the cable. Mini45 has an ultra-high-strength body with a maximum permissible overload value of 5.7 to 25.3 times the rated range. Mini45 also has not only a high signal-to-noise ratio (75 times that of conventional metal strain gages), but also a near-zero noise distortion of the amplified signal.

The MYO Armband (Fig. 2.8) is an electromyography (EMG) signal acquisition device manufactured by Thalemic Labs. The user can carry the MYO Armband on the arm to achieve wireless human–computer interaction.

Figure 2.8 MYO Armband.

Eight surface EMG (sEMG) sensors are built into the MYO Armband to capture the sEMG signal on the surface of the arm, while a nine-axis IMU sensor is also built in to acquire acceleration and gyroscope signals. When the user wears the MYO Armband to perform different gestures, the high-sensitivity sensor processes the captured signal and transmits the three types of signals through low-power Bluetooth, which can recognize different gestures. Because everyone's muscular system and nervous system are different, sensors need to use the electrical drive of the operator's arm muscles to create information. By using the pre-calibration process, the MYO Armband can more accurately identify gesture movements.

The MYO Armband can be used in our daily life to control screens and play music through gestures, and it can also be applied to remote control smart devices and medical aids. In addition, the MYO Armband has also been applied to robotics research, such as the teleoperation of mobile robots [11,12].

2.2. Software systems

2.2.1 OpenHaptics toolkit

The OpenHaptics toolkit is a host computer programming interface provided by Geomagic for its haptic devices. The OpenHaptics toolkit enables users to go beyond working with a 2D mouse in their applications to interact with and manipulating objects realistically and intuitively [13]. It allows users to feel objects in a virtual 3D scene, making skills easier to learn, in a way provides true 3D navigation and direct interaction. OpenHaptics enables software developers to add haptics and true 3D navigation to a broad range of applications including 3D design and modeling, medicine, games, entertainment, visualization, and simulation. This haptics toolkit is patterned after the OpenGL API, making it familiar to graphics programmers and facilitating integration with OpenGL applications. Using the OpenHaptics toolkit, developers can leverage existing OpenGL code for specifying geometry and supplement it with OpenHaptics commands to simulate haptic material properties such as friction and stiffness. The extensible architecture enables developers to add functionality to support new types of shapes. It is also designed to integrate third-party libraries such as physics/dynamics and collision detection engines. The OpenHaptics toolkit supports the range of 3D Systems PHANTOM devices, from the more compact Touch device to the larger PHANTOM Premium devices. The OpenHaptics toolkit supports Microsoft Windows and Linux.

This toolkit includes the Haptic Device API (HDAPI), the Haptic Library API (HLAPI), utilities, PHANTOM Device Drivers (PDDs), and source code examples. The HDAPI provides low-level access to the haptic device, enables haptic programmers to render forces directly, offers control over configuring the runtime behavior of the drivers, and provides convenient utility features and debugging aids. The HLAPI provides high-level haptic rendering and is designed to be familiar to OpenGL API programmers. It allows significant reuse of existing OpenGL code and greatly simplifies synchronization of the haptics and graphics threads. The Device Drivers support all 3D Systems' Touch and Phantom haptic devices.

As for haptic devices, the 3D Systems product line of haptic devices enables users to touch and manipulate virtual objects. Different models in this industry-leading product line meet the varying needs of commercial software developers, academic and commercial researchers, and product designers. The PHANTOM Premium models are high-precision instruments and, within the PHANTOM product line, provide the largest workspaces and highest forces, and some offer six-degrees-of-freedom output capabilities. The Geomagic Touch and Touch X devices offer affordable desktop solutions.

2.2.2 MATLAB® Robotics Toolbox

MATLAB is a software platform that links computing, programming, and interfacing and is commonly used as simulation software in many disciplines. MATLAB is extremely inclusive and contains many different types of toolboxes for researchers to solve problems in various fields. Especially, vectors and matrices are the basic data types of MATLAB, which are quite suitable for solving problems related to robotics.

The MATLAB Robotics Toolbox [14] is a MATLAB-based robotics toolbox developed and managed by Peter Corke's team at the Scientific and Industrial Research Organisation in Australia. It can be downloaded for free from the website provided by Peter Corke (https://petercorke.com/toolboxes/robotics-toolbox/). The toolbox provides a library of functions and examples of several common robots. The function library consists of functions such as flush transformations, 3D modeling, forward and inverse kinematics solution, dynamics solution, and trajectory planning. The robot model can be easily and quickly set up in the MATLAB environment using program functions, and then the model can be directly simulated. The example objects in the toolbox include the classic PUMA560 robot, the KUKA robot, and the Baxter robot. With the help of MATLAB simulation, the operation of each joint can be studied dynamically. Therefore, it is very convenient to use the Robotics Toolbox for robot modeling, trajectory planning, control, and visualization.

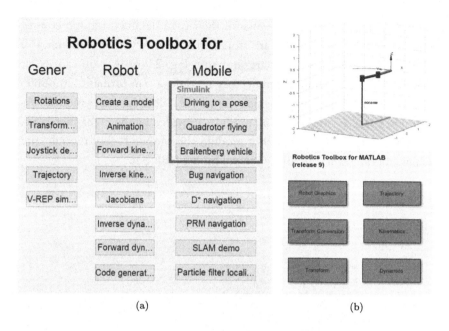

Figure 2.9 Robotics Toolbox.

The start interface of the Robotics Toolbox is shown in Fig. 2.9(a). To simulate a robot through the MATLAB Robotics Toolbox, a model of the robot object must be constructed [15–17]. The MATLAB Robotics Toolbox builds the robot model with the help of the robot's Denavit–Hartenberg (D-H) parameters, which describe the coordinate transformations between the robot's neighboring links and the Link and SeriaLink functions in the toolbox. The Link function is used to create individual bars, and the SeriaLink function is used to connect the individual bars to each other. As shown in Fig. 2.9(b), the plot function in the toolbox can also display the created model. In addition, the Robotics Toolbox provides transformation tools for handling different data types. For example, quaternions, representing 3D positions and orientations, can use homogeneous transformation through conversion tools to complete the corresponding transformation conveniently and quickly.

Many functions in the Robotics Toolbox can be used within Simulink to perform block diagram design. Simulink is a companion product to MATLAB that provides dynamic system simulation based on a block diagram modeling language. The wrapper module of the toolbox function can describe nonlinear robotic systems in block diagram form, allowing users to study the closed-loop performance of their own designed robotic control systems. The robot modules are stored in toolbox\rvctool\simulink. All robot modules in the toolbox are shown in Fig. 2.9(b). As you can see, it includes a graphics module, a trajectory module, a transformation module, a kinematics module, a transformation module, and a dynamics module. As shown in Fig. 2.10, the Simulink modules provided by the Dynamics module include the Articulated Robot module, the Mobile Robot module, the Vehicle module, and the Joint Control model [18].

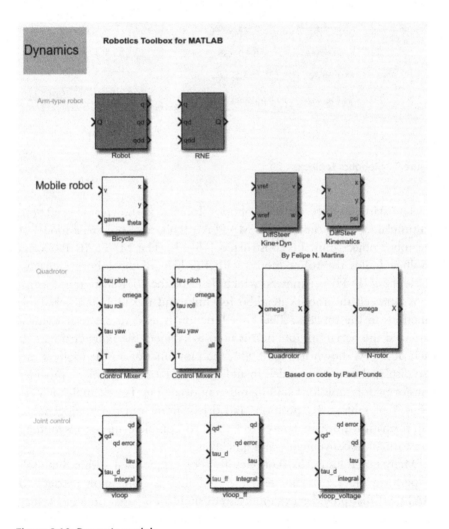

Figure 2.10 Dynamic module.

This book covers the study of the robot model including flush coordinate transformation, kinematic solution, construction of model objects, and tracking of each motion information in its trajectory planning through the common methods of the Robotics Toolbox.

2.2.3 Robot operating system

With the rapid development and complexity of the robotics field, the demand for code reusability and modularity is growing, but the existing open source robotics systems cannot meet the demand well. In 2010, Willow Garage released ROS, an open source robotics operating system, which quickly led to a boom in the robotics research field. The ROS system was the product of a collaboration between a project at Stanford University's Artificial Intelligence Lab and the Personal Robots Program of robotics company Willow Garage in 2007.

The official website of ROS (http://wiki.ros.org) provides a brief description of ROS: "ROS provides a series of libraries and tools to help software developers create robotic applications. It also provides hardware abstraction, device drivers, function libraries, visualization tools, messaging, and package management." However, for abecedarians, the above brief introduction will not give them a deeper understanding of ROS features, so the following is a deeper explanation [19].

ROS is an open source system that is a post-operating system, or sub-operating system, for robots. It provides functions similar to those provided by an operating system, including hardware abstraction description, underlying driver management, execution of common functions, messaging between programs, and program distribution package management. It also provides a number of tool programs and libraries for acquiring, building, writing, and running programs for multimachine integration.

The primary design goal of ROS is to increase code reuse in the field of robotics development. ROS is a distributed processing framework (a.k.a. Nodes). This allows executables to be designed individually and loosely coupled at runtime. These processes can be encapsulated in Packages and Stacks for easy sharing and distribution. ROS also supports a federated system of code libraries, allowing collaboration to be distributed as well [20]. The design from the file system level to the community level makes it possible to make independent decisions about development and implementation. All of the above capabilities can be implemented using the ROS-based tools.

The ROS operational architecture is a processing architecture that uses ROS communication modules to implement a loosely coupled network connection between modules on P2P. It performs several types of communication, including service-based synchronous remote procedure call (RPC) communication, topic-based asynchronous data flow communication, and data storage on parameter servers. But ROS itself is not real-time. The main features of ROS can be summarized as follows [21].

(1) Peer-to-peer design

A system using ROS consists of a series of processes that exist in multiple different hosts and are connected through an end-to-end topology during operation. Although those software frameworks based on central servers can also realize multiple processes and multiple hosts, in these frameworks, problems may occur with the central data server when the computers are connected through different networks.

The peer-to-peer design of ROS and mechanisms such as service and node managers can spread the pressure of real-time computation caused by functions such as computer vision and speech recognition, and can adapt to the challenges encountered by multiple robots.

(2) Multilingual support

When it comes to coding, many programmers have a preference for certain programming languages. These preferences are the result of individual differences in programming time, debugging effectiveness, syntax, execution efficiency, and various technical and cultural differences in each language. To address these issues, ROS has been designed as a language-neutral framework structure. ROS supports many different languages, such as C++, Python, Octave, and LISP, and also contains multiple interface implementations for other languages.

The special features of ROS are mainly found at the message communication layer, not at a deeper level. End-to-end connectivity and configuration are implemented by the XML-RPC mechanism, which also contains reasonable implementation descriptions for most major languages. It is expected that ROS could be implemented more naturally using a variety of languages and more in line with the syntactic conventions of each language, rather than providing implementation interfaces to a variety of other languages based on C. However, it is convenient to support more new languages with already existing libraries in some cases; for example, Octave's client is implemented through a C++ wrapper library. To support language crossover, ROS makes use of a simple, language-independent interface definition language to describe the messaging between modules. A short text is used to describe the structure of each message, which also allows for message composition.

The code compiler for each language would then generate similar target files for this language, which can be automatically implemented in successive parallel by ROS during message passing and receiving. This saves important programming time and avoids errors. The previous three-line interface definition file is automatically expanded into 137 lines of C++ code, 96 lines of Python code, 81 lines of Lisp code, and 99 lines of Octave code. Because messages are automatically generated from various simple text files, it is easy to enumerate new message types. At the time of coding, the known ROS-based code base contains over 400 message types that transmit data from sensors that make objects detect their surroundings. The final result is that the language-independent message processing allows multiple languages to be freely mixed and matched for usage.

(3) Streamlining and integration

Most of the existing robotics software projects contain drivers and algorithms that can be reused outside of the project. Unfortunately, the middle layer of most codes is very confusing for multiple reasons that it is difficult to extract its functionality and to apply them from the prototype to other aspects.

In response to this trend, it was encouraged to gradually develop all drivers and algorithms into separate libraries with no reliance on the ROS. The system built by ROS is modular, with the code in each module compiled separately. The CMake tool used for compilation makes it easy to implement the concept of streamlining. ROS encapsulates complex code in libraries, so that simple code can be ported and reused beyond the prototype with just a few small applications created.

ROS leverages many codes from open source projects that already exist today, such as code from the Driver, Motion Control, and Simulation of the Player project, code from the vision algorithms of OpenCV, planning algorithms from OpenRAVE, and many others. In each instance, ROS is used to display a wide variety of configuration options and to communicate data to and from the software, as well as to perform minor packaging and make changes to them. ROS can be continually upgraded from community maintenance, including upgrading the ROS source code from other software repositories and application patches.

(4) Extensive toolkit

To manage the complex ROS software framework, a large number of widgets are utilized to compile and run a wide variety of ROS components, thus designed as a kernel, rather than building a huge development and runtime environment.

These tools perform a variety of tasks, such as organizing the structure of the source code, obtaining and setting configuration parameters, visualizing end-to-end topology connections, measuring band usage widths, vividly depicting information data, and automatically generating documentation. Although the core services of the logger like the global clock and controller modules have been tested, it is still hoped that all the code will be modularized. It is believed that the loss in efficiency is far from being compensated by the complexity of stability and management.

(5) Free and open source

All source codes of ROS are publicly released. We believe this will certainly facilitate debugging and bug correction at all levels of the ROS software. While non-open source programs such as Microsoft Robotics Studio and Webots have many commendable attributes, we believe there is no substitute for an open source platform. This is especially true when hardware and software at all levels are designed and debugged simultaneously.

ROS follows the BSD license in a distributed relationship, which means that it allows a variety of commercial and non-commercial projects to be developed, and ROS passes data through an internally processed communication system that does not require modules to be linked together in the same executable. Thus, systems built with ROS can make good use of their rich components: individual modules can contain software protected by various protocols ranging from GPL to BSD, but some of the "contaminants" of the license are completely eliminated in the module decomposition.

2.2.4 Gazebo

Gazebo is one of the most widely used robotics simulation development platforms in the world [22]. Its interface is shown in Fig. 2.11. Gazebo can be used to simulate robots in a 3D environment, including robot–obstacle interaction and physical properties of collisions. Through Gazebo's 3D simulation environment, one can watch the robot's operation in the environment, which greatly facilitates the research of robot navigation algorithms. The simulation environment provides an excellent simulation of

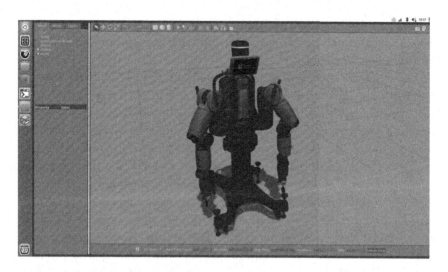

Figure 2.11 Gazebo.

the robot motion process and the dynamics involved, various sensors, and robot motion planning and control [23].

Gazebo is an open source 3D robot dynamics simulation program that simulates articulated robots in complex and realistic environments. It allows efficient simulation of robot path planning, positioning, mapping, and other navigation algorithms in a 3D environment. It can be used with ROS for complete robot simulation or on its own.

Gazebo supports 3D-based rendering and physical simulation. It can simulate various sensors, including sonar, scanning laser rangefinders, GPS, and monocular and stereo cameras, and various robots, such as PR2, Pioneer2DX, and Pioneer2AT. It is highly configurable and scalable. Users can design a variety of complex heterogeneous robot models in a realistic virtual environment and dynamically load these models at runtime. The robot's operation can be further enhanced by programming according to user needs, and controller programs can be written in C, C++, etc. Gazebo is a robot simulator in the 3D world. It is equipped with an advanced plugin interface that can be used to simulate sensor feedback and object interactions. Gazebo uses SDF (Gazebo's models are files with the URDF or SDF suffix) to model robots and other objects [24]. In addition, Gazebo has other established standard models that can be used, and users can create models to suit their needs. For example, Gazebo uses its URDF to model Baxter. Controller plugins such as block lasers, point lasers, and cameras are used to simulate Baxter's sonar devices, infrared devices, and cameras. The simulator can be integrated with rviz and moveit and used for real robots.

Gazebo is the default robotics simulation software used in ROS. Gazebo and ROS are separate projects, i.e., their developers are not the same, but there is a package related to Gazebo in the official ROS repository (ros-indigo-gazebo-ros). This is a package maintained by the Gazebo developers themselves and it contains plugins that interface with ROS and Gazebo. These plugins can be connected to objects in the emulation software scenario and provide simple ROS communication methods; for example, Gazebo publishes and subscribes to themes and services. Wrapping Gazebo as a ROS node also allows it to be easily integrated into the default ROS method of running large and complex systems (called boot files). Gazebo has a clear advantage: although they are separate projects, each version of Gazebo has been developed with the ROS version in mind, so that Gazebo can keep up with the speed of ROS updates. Gazebo has integrated some ROS-specific features such as services, topics, subscriptions, and publications. Gazebo and ROS already have a large number of community-developed plugins and code [25].

Generally, the typical uses of Gazebo include:
• testing robotics algorithms,
• designing robots,
• regression testing with realistic scenarios.

The key features of Gazebo include the following:
• multiple physics engines,
• a rich library of robot models and environments,
• a wide variety of sensors,
• it is easy to program and has a simple graphical interface.

2.2.5 CoppeliaSim

CoppeliaSim (Fig. 2.12), formerly known as V-rep, is a very flexible and scalable dynamics simulation software. Its distributed control system allows it to integrate a number of very powerful motion simulation libraries, such as the remote API and ROS interfaces. These simulation libraries can be used to simulate the kinematic dynamics of the entire physical structure of the robot and the overall robot motion control system, thus making it possible to simulate the motion processes required for our experiments more accurately and more easily. Moreover, CoppeliaSim has a very rich API function, through which we can easily implement a variety of functions

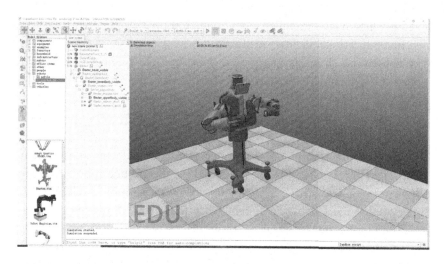

Figure 2.12 CoppeliaSim.

in robot control. So, CoppeliaSim can be used in a wide range of applications such as remote video monitoring, safety inspection, robot hardware control, robot teaching, mechanical system kinematics and dynamics verification, robot control algorithm development and parameter adjustment, the fully automated simulation of machining processes in factories, and the presentation of corporate products, among a range of other areas. The common computational modules in CoppeliaSim are dynamic simulation engines, forward/inverse kinematics tools, collision detection libraries, vision sensor simulation, path planning, and GUI development tools. Among these more commonly used calculation modules, the dynamic simulation engine is mainly used to perform physical simulations of dynamics calculations. There are four engines to choose (Bullet Physics, Open Dynamics Engine, Vortex Studio, and Newton Dynamics), which are mainly used for dynamics simulation. Users can choose the appropriate simulation engine according to the actual simulation needs. In addition, there are many 3D models of common robots built in CoppeliaSim, which makes it possible to master CoppeliaSim more quickly. In general, the key features of CoppeliaSim include the following [26]:

- CoppeliaSim provides a unified framework that combines a number of very powerful internal and external libraries for simulation.
- CoppeliaSim is highly extensible and provides developers with a powerful API.

- CoppeliaSim is a cross-platform and open source program which offers a free educational license certificate.
- CoppeliaSim supports programming in multiple languages, including Java, C/C++, LUA, Python, MATLAB, Urbi, and Octave.

2.3. Conclusion

This chapter introduces a number of robot platforms and relevant devices used throughout this book, including the mobile robot, the humanoid robot platform Baxter, and the collaborative robot KUKA LBR iiwa, as well as the Geomagic Touch X haptic device. Meanwhile, a number of useful software toolkits in robot simulation are also introduced in this chapter, such as the OpenHaptics toolkit for haptic devices, the MATLAB Robotics Toolbox, Gazebo, and CoppeliaSim. These devices and toolkits are increasing used in the study of robotics, as they provide ideal means for the study, design, and testing of robotic technologies.

References

[1] Mulun Wu, Yanbin Xu, Chenguang Yang, Ying Feng, Omnidirectional mobile robot control based on mixed reality and sEMG signals, in: 2018 Chinese Automation Congress (CAC), 2018, pp. 1867–1872.

[2] Chenguang Yang, Xinyu Wang, Long Cheng, Hongbin Ma, Neural-learning-based telerobot control with guaranteed performance, IEEE Transactions on Cybernetics (2016).

[3] C. Yang, C. Chen, W. He, R. Cui, Z. Li, Robot learning system based on adaptive neural control and dynamic movement primitives, IEEE Transactions on Neural Networks and Learning Systems 30 (3) (2019) 777–787.

[4] C. Yang, Neural learning enhanced teleoperation control of Baxter robot using IMU based motion capture, in: 2016 22nd International Conference on Automation and Computing (ICAC), 2016.

[5] Chunxu Li, Chenguang Yang, Jian Wan, Andy S.K. Annamalai, Angelo Cangelosi, Teleoperation control of Baxter robot using Kalman filter-based sensor fusion, Systems Science & Control Engineering 5 (1) (2017) 156–167.

[6] C. Yang, Y. Jiang, Z. Li, W. He, C.Y. Su, Neural control of bimanual robots with guaranteed global stability and motion precision, IEEE Transactions on Industrial Informatics (2016) 1.

[7] C. Li, C. Yang, Z. Ju, Ask Annamalai, An enhanced teaching interface for a robot using DMP and GMR, International Journal of Intelligent Robotics & Applications 2 (1) (2018) 110–121.

[8] C. Yang, J. Luo, Y. Pan, Z. Liu, C.Y. Su, Personalized variable gain control with tremor attenuation for robot teleoperation, IEEE Transactions on Systems, Man, and Cybernetics: Systems (2017).

[9] C. Li, C. Yang, P. Liang, A. Cangelosi, W. Jian, Development of Kinect based teleoperation of Nao robot, in: 2016 International Conference on Advanced Robotics and Mechatronics (ICARM), 2016.

[10] Huifeng Lin, Chenguang Yang, Silu Chen, Ning Wang, Min Wang, Zhaojie Ju, Structure modelling of the human body using FGMM, in: 2017 IEEE International Conference on Cybernetics and Intelligent Systems (CIS) and IEEE Conference on Robotics, Automation and Mechatronics (RAM), 2017, pp. 809–814.

[11] J. Luo, Z. Lin, Y. Li, C. Yang, A teleoperation framework for mobile robots based on shared control, in: 2020 IEEE International Conference on Robotics and Automation (ICRA), 31 May–31 August 2020, 2020.

[12] Y. Xu, C. Yang, X. Liu, Z. Li, A teleoperated shared control scheme for mobile robot based sEMG, in: 2018 3rd International Conference on Advanced Robotics and Mechatronics (ICARM), 2018.

[13] https://www.3dsystems.com/haptics-devices/openhaptics/.

[14] Alessandro Gasparetto, V. Zanotto, A new method for smooth trajectory planning of robot manipulators, Mechanism and Machine Theory 42 (4) (2007) 455–471.

[15] A. Ismael F. Vaz, Edite M.G.P. Fernandes, M. Paula S.F. Gomes, Robot trajectory planning with semi-infinite programming, European Journal of Operational Research 153 (3) (2004) 607–617.

[16] F.L. Lewis, C. Abdallah, D.M. Dawson, Control of robot manipulators, 1993.

[17] Haruhiko Asada, J.-J.E. Slotine, Robot Analysis and Control, John Wiley & Sons, 1991.

[18] Peter Corke, Matlab toolboxes: Robotics and vision for students and teachers, IEEE Robotics & Automation Magazine 14 (4) (2007) 16–17.

[19] http://wiki.ros.org.

[20] Aaron Martinez, Enrique Fernández, Learning ROS for Robotics Programming, Packt Publishing Ltd, 2013.

[21] P. Abbeel, A. Albu-Schaffer, A. Ames, F. Amigoni, H. Andreasson, P. Asaro, C. Balaguer, J. Baltes, E. Bastianelli, S. Behnke, et al., 2015 index, IEEE Robotics & Automation Magazine 22 (2015).

[22] Jens Kober, Learning motor skills: From algorithms to robot experiments, it – Information Technology 56 (3) (2014) 141–146.

[23] http://www.gazebosim.org/.

[24] Nathan Koenig, Andrew Howard, Design and use paradigms for gazebo, an open-source multi-robot simulator, in: 2004 IEEE/RSJ International Conference on Intelligent Robots and Systems (IROS) (IEEE Cat. No. 04CH37566), vol. 3, IEEE, 2004, pp. 2149–2154.

[25] Johannes Meyer, Alexander Sendobry, Stefan Kohlbrecher, Uwe Klingauf, Oskar Von Stryk, Comprehensive simulation of quadrotor UAVs using ROS and Gazebo, in: International Conference on Simulation, Modeling, and Programming for Autonomous Robots, Springer, 2012, pp. 400–411.

[26] https://coppeliarobotics.com/helpfiles/.

CHAPTER THREE

Uncertainties compensation-based teleoperation control

3.1. Introduction

Teleoperation technology brings benefits to humans. The performance of a teleoperation system is greatly influenced by system uncertainties, which are caused by time delays in the communication channel of the system [1,2] and the dynamics in the actual environment [3].

3.1.1 Wave variable method

For time delays in communication channels, it is noted that the system is destabilized by the time delays in a closed-loop system [4,5]. Moreover, the stability and transparency of the system are sensitive to the communication time delay issue [6]. To guarantee the stable performance of the teleoperation system, many solutions have been introduced in the literature such as the scattering theory network approach and the passivity method [7,8].

In these methods, the wave variable approach is significant because of its construction and passivity [9]. Huang et al. [10] proposed a method based on the wave variable to guarantee stability and tracking performance of the position and force for dual-leader dual-follower (DLDF) systems. A radial basis function neural network (RBFNN) control method with wave variables was developed to reduce the influences of time delays and dynamics uncertainties [11]. Yuan et al. [12] developed a force observer with dynamic gain to collect the force reflection in prescribed performance functions and to obtain satisfactory manipulation performance. Soyguder and Abut [13] proposed a novel control method with time delays to ensure stable performance of position tracking of a haptic industrial robot. A new wave variable method to strengthen the performance of haptic feedback, reduce bias, and guarantee steady-state position tracking in teleoperation was proposed [14]. Additionally, in Ref. [15], an ideal method was proposed to augment the wave and a wave variable method was proposed to guarantee the tracking performance of the follower.

Stability and transparency are both main objectives in teleoperation control. A general four-channel teleoperation architecture can achieve ideal transparency [16]. Since the stability and transparency of the system are sensitive to the communication time delay issue [6], there have been many studies about the four-channel architecture in the wave variable method. Sun et al. [17,18] proposed a new four-channel method with a modified wave variable controller to improve the stability and transparency of the teleoperation systems. An approach with a four-channel structure of Lawrence was proposed to guarantee the passivity through the wave variable and the absolute stability of the system in terms of position and force [19]. Pitakwatchara et al. [20] developed a novel wave variable method with a wave correction scheme to handle the problem of motion incongruity in task space for the teleoperation system. In Ref. [21], a novel wave variable method with four channels is presented to achieve stable tracking in position and force in bilateral teleoperation.

3.1.2 Neural learning control

In a practical robot system, the existence of nonlinear and uncertainties in dynamic parameters is general, which can degrade the performance and even the stability of the robot system. Thus, studies on controlling the uncertainties in robot systems are important. Adaptive control can eliminate the uncertain influence of the robot system as much as possible by estimating and correcting the unknown parameters of the system online. In Ref. [22], adaptive fuzzy control was used for studying an uncertain nonlinear system. In Ref. [23], an adaptive control structure was used in time-varying delays and uncertainties in teleoperation.

In recent years, the applications of neural networks (NNs) in robot system control have become popular, because NNs can simulate complex nonlinear and uncertain functions [24,25]. RBFNN is a highly effective method that has been widely used in the control design of uncertain robot systems. In Ref. [26], the adaptive neural network method was applied to control uncertain marine vessel systems. To compensate for the nonlinear problems that the standard PD controller cannot deal with, the authors in Ref. [27] developed the RBFNN as a compensator. In Ref. [28], the RBFNN is used to learn the behavior of the robot, and in Ref. [29], RBFNN is used to simultaneously compensate for the effects caused by dynamics uncertainties in the bilateral teleoperation system. In addition, the authors further proposed an NN with a four-channel method to ensure the passivity and high transparency of the systems and to estimate the dynamic uncertainties of the systems [30].

In this chapter, we will introduce the RBFNN that is used as a dynamics compensator in bilateral teleoperation systems. The wave variable method to deal with time delay in bilateral teleoperation will be discussed, for both the two-channel situation and the four-channel situation.

3.2. System description

3.2.1 Dynamics of teleoperation systems

As shown in Fig. 3.1, the dynamics of a bilateral teleoperation system can be represented as

$$m_l(x_l)\ddot{x}_l + c_l(x_l, \dot{x}_l) + g_l(x_l) = f_h - f_l, \tag{3.1}$$

$$m_f(x_f)\ddot{x}_f + c_f(x_f, \dot{x}_f) + g_f(x_f) = f_f - f_e, \tag{3.2}$$

where subscripts "l" and "f" are used to indicate leader and follower, respectively, m_l and m_f indicate the inertia matrices for the teleoperation system, c_l and c_f are the Coriolis and centrifugal force matrices, respectively, g_l and g_f are the gravitational forces matrix of the system, x_i is the position, $\dot{x}_i$ is the velocity, $\ddot{x}_i$ is the acceleration ($i = l, f$), f_l indicates the output force of the leader device, and f_h is the applied force to the leader device by the human operator. Similarly, f_f represents the output force of the follower robot and f_e represent the interaction force between the follower robot and the remote environment. T is delay time. u_i and v_i ($i = l, f$) are the wave variables.

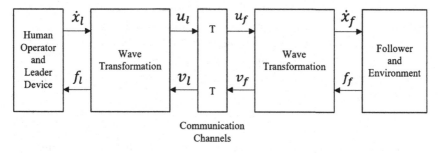

Figure 3.1 A typical bilateral teleoperation system.

· Property 1. $m_i(x_i)$ is a symmetric positive definite matrix of the teleoperation system.

· Property 2. $z^T(\dot{m}_i(x_i) - 2c_i(x_i, \dot{x}_i))z = 0, \forall z \in R^n$.
· Property 3. m_i is a bounded term. g_i is also a bounded term. It satisfies $\forall x_i, \dot{x}_i \in R^n, \exists K_{ci} \in R > 0$ according to c_i, so that $\|c_i(x_i, \dot{x}_i)\| <= K_{ci}|\dot{x}_i|$.

Similarly, in joint space, the dynamics can be represented by the joint variable $(q_i, \dot{q}_i, \ddot{q}_i)$:

$$m_l(q_l)\ddot{q}_l + c_l(q_l, \dot{q}_l) + g_l(q_l) = J^T f_h - \tau_l, \tag{3.3}$$

$$m_f(q_f)\ddot{q}_f + c_f(q_f, \dot{q}_f) + g_f(q_f) = \tau_f - J^T f_e, \tag{3.4}$$

where J^T represents the Jacobian matrix, converting force in Cartesian space to torque in joint space.

3.2.2 Position-to-position control in teleoperation systems

To realize the position–position control strategy, the tip position of the leader device is treated as the desired position of the follower robot. One solution is to directly match the joint angles of the leader device and the follower robot one by one. Although this solution is intuitive and simple, for the follower manipulator with redundant configuration, this solution cannot fully utilize the redundant performance of these kinds of manipulators. Thus, it is necessary to use the inverse kinematics from the leader device to the follower robot in Cartesian space.

A closed-loop inverse kinematics (CLIK) algorithm is employed. By using the Jacobian, the relationship between joint angles and tip velocity can be described as follows:

$$\dot{q} = J^+(q)\dot{x}, \tag{3.5}$$

where J^+ is the pseudo-inverse of the Jacobian matrix. Replacing the task space velocity vector $\dot{x}$ with $\dot{x} = K(x_d - x)$, numerical drift in the task space can be avoided. K is a positive definite matrix to ensure convergence of the position error [31,32]. To reduce the computation of matrix inversion, the Jacobian matrix $J_T(q)$ is used to replace J^+. Thus, the CLIK algorithm can be represented as

$$\dot{q} = KJ^T(q)(x_d - x). \tag{3.6}$$

This algorithm can avoid numerical instability when moving singular points, due to the lack of the required pseudo-inverse of the Jacobian matrix. Calculation details and stability proof of this solution are shown in Refs. [33–35]. The block diagram of the CLIK algorithm is shown in Fig. 3.2.

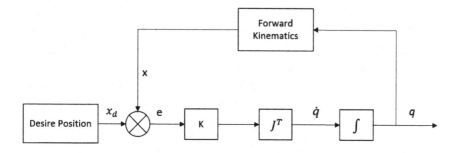

Figure 3.2 Block diagram of the CLIK algorithm.

3.2.3 Four-channel control in teleoperation systems

Fig. 3.3 presents the general four-channel teleoperation architecture. The dynamics of a four-channel architecture can be represented as follows:

$$F_h^* = F_h + Z_h V_l', \tag{3.7}$$

$$F_e^* = F_e + Z_e V_f', \tag{3.8}$$

$$Z_{cl} V_l' + C_4 V_f' = (1 + C_6) F_h - C_2 F_e, \tag{3.9}$$

$$C_1 V_l' - Z_{ce} V_f' = (1 + C_5) F_e - C_3 F_h, \tag{3.10}$$

where $Z_{cl} = Z_l + C_l$, $Z_{ce} = Z_f + C_f$, F_h^* and F_e^* indicate the applied force of the human operator and the force exerted on the environment, F_h is the interaction force of the human operator and the leader device, F_e represents

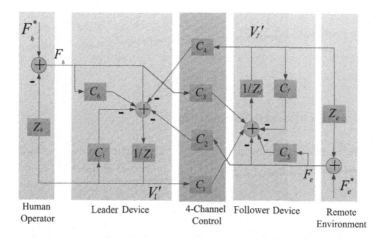

Figure 3.3 The general four-channel control teleoperation system.

the interaction force of the environment and the follower device, Z_h is the impedance of the human operator, Z_e is the impedance of the environment, V_l' and V_f' indicate the velocities of the leader device and the follower device, respectively, C_1–C_6 are the enrollment parameters to impact the transparency performance of the teleoperation system, and Z_i and C_i ($i = l, f$) are the impedance parameters and the local position control parameters of the leader device and the follower device, respectively.

3.3. Neural learning control

3.3.1 Principle of RBFNN

RBFNN is an artificial neural network (NN) that takes the radial basis function (RBF) as an activation function in the hidden layer. RBFNN can map the nonlinearly separable vectors in low dimension into a high-dimensional space, which makes them linearly separable. RBNN has many useful applications, including approximating nonlinear functions, time series prediction, classification, and system control.

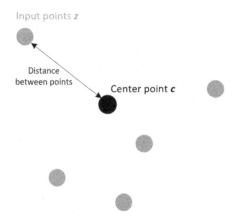

Figure 3.4 Principle of the radial basis function.

The RBF is a kind of scalar function symmetrical along the radial direction. As shown in Fig. 3.4, the RBF is expressed as follows:

$$\varphi(z) = \varphi \|z - c\|. \tag{3.11}$$

The value of the RBF is only related to the distance between the point z and a certain origin c. If any function satisfies the above formula, it can be called the RBF. Since the Gaussian function has the advantages of smoothness and differentiability at any order, it has become a commonly used RBF:

$$\varphi(z) = exp(-\frac{\|z - c\|^2}{\sigma^2}). \qquad (3.12)$$

The structure of a neuron in RBFNN is shown in Fig. 3.5.

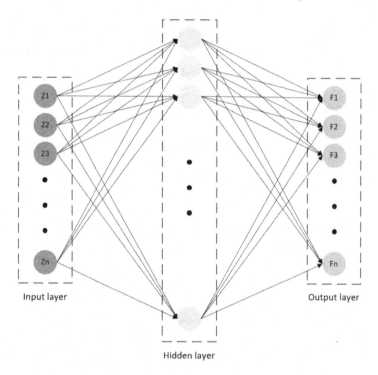

Figure 3.5 Structure of RBFNN.

The RBFNN is a three-layer NN with a single hidden layer. The first layer is the input layer, which is composed of signal input nodes $z = [z_1, z_2, \cdots, z_n]^T$. The hidden layer is composed of RBF neurons. And the output layer is the output vector, which is a linear combination of the hidden layer outputs:

$$F(z) = \hat{W}^T \varphi(z). \qquad (3.13)$$

Here, $\hat{W}^T \in R^{N*n}$ is the weight matrix that connects the hidden layer and the output layer, N and n represent the dimension of the hidden layer and output layer, respectively, and in the dynamic compensation case, it corresponds to the degree of freedom of the follower robot arm.

3.3.2 Nonlinear dynamic model

Considering the dynamics of the follower robot in joint space in Eq. (3.4), a conventional PD controller can be designed:

$$\tau_f = -K_f e_f - D_f \dot{e}_f, \tag{3.14}$$

where $e_f = q_f - \dot{q}_f$ is the tracking error and K_f and D_f are the symmetric positive definite matrices for the joint angle and angular velocity gains, respectively.

By defining the generalized tracking error [11],

$$e_{vf} = \dot{e}_f + K_{f1} e_f K_{f1} = D_f^{-1} e_{vf}, \tag{3.15}$$

we have

$$\tau_f = -D_f e_{vf}. \tag{3.16}$$

Defining $q_v = \dot{q}_d - K_{f1} e_f$, the dynamics of the follower robot Eq. (3.4) can be rewritten as

$$m_f \dot{e}_{vf} + c_f e_{vf} + D_f e_{vf} = -J^T f_e - g_f + m_f \dot{q}_v - c_f q_v. \tag{3.17}$$

Then, the uncertain nonlinear dynamics can be described as

$$F(z) = -J^T f_e - g_f + m_f \dot{q}_v - c_f q_v, \tag{3.18}$$

$$m_f \dot{e}_{vf} + c_f e_{vf} + D_f e_{vf} = F(z). \tag{3.19}$$

From the above equation, we can observe that the steady-state error will not be zero because $F(z) \neq 0$, so dynamic compensation is needed.

3.3.3 Approaching nonlinear dynamic model with RBFNN

RBFNN can be used to approximate the dynamics of the follower robot by using the output vector $\hat{F}(z)$ to approximate the uncertain nonlinear function $F(z)$ in Eq. (3.19). The following lemmas are given [11].

Lemma 1. *The input vector of RBFNN is $x \in X$, where X is a compact subset.*

Lemma 2. *Given a positive constant ϵ_0 and a continuous function $F: z \rightarrow R^n$, there exists a weight matrix $W^* \in R^{N*n}$ making the output vector of the RBFNN $\hat{F}(z)$ satisfy*

$$\max_{z \in X} \|\hat{F}(z) - F(z)\| <= \epsilon. \tag{3.20}$$

Lemma 3. *The output vector of an RBFNN on its arguments z, $\hat{W}$ is continuous.*

Therefore, Eq. (3.19) can be rewritten as follows:

$$m_f \dot{e}_{vf} + c_f e_{vf} + D_f(e_{vf}) = F(z) - \hat{F}(z). \tag{3.21}$$

According to the properties of the RBFNN shown before, Eq. (3.21) could be rewritten as

$$m_f \dot{e}_{vf} = -(c_f e_{vf} + D_f(e_{vf})) + (F(z) - \hat{W}^T \varphi(z)). \tag{3.22}$$

Then the update law of the weight matrix can be obtained with the Lyapunov method [11]:

$$\dot{\hat{W}} = -Q^{-1}\varphi(z)e_{vf}^T, \tag{3.23}$$

where Q is a symmetric positive definite matrix.

3.4. Wave variable method

3.4.1 Wave variable correction-based method

(1) Traditional wave variable method

Figure 3.6 The general wave variable method.

Fig. 3.6 displays the general wave variable method for teleoperation systems. The wave transformation shown in the figure can complete the conversion between the velocity $\dot{x}$, force F, and the power variables U, V are as follows:

$$U_l = \frac{b}{\sqrt{2b}}\dot{x}_{cl} + \frac{1}{\sqrt{2b}}\dot{f}_{cl},$$

(3.24)

$$V_l = \frac{b}{\sqrt{2b}}\dot{x}_{cl} - \frac{1}{\sqrt{2b}}\dot{f}_{cl},$$

(3.25)

$$U_f = \frac{b}{\sqrt{2b}}\dot{x}_{cf} + \frac{1}{\sqrt{2b}}\dot{f}_{cf},$$

(3.26)

$$V_f = \frac{b}{\sqrt{2b}}\dot{x}_{cf} - \frac{1}{\sqrt{2b}}\dot{f}_{cf},$$

(3.27)

where the wave impedance $b > 0$.

Considering the time delays T_1 and T_2 existing in the communication channels, the wave variables could be represented as follows:

$$U_f = U_l(t - T_1(t)),$$

(3.28)

$$V_l = V_f(t - T_2(t)).$$

(3.29)

From Fig. 3.6, the relationship between the wave variables and the power variables is as follows:

$$\dot{x}_l = \frac{(U_l + V_l)}{\sqrt{2b}},$$

(3.30)

$$\dot{F}_l = \frac{(U_l - V_l)}{\sqrt{2b}},$$

(3.31)

$$\dot{x}_f = \frac{(U_f + V_f)}{\sqrt{2b}},$$

(3.32)

$$\dot{F}_f = \frac{(U_f - V_f)}{\sqrt{2b}}.$$

(3.33)

Then the flowing power in the communication channels could be calculated:

$$P(t) = \dot{x}_l^T F_l - \dot{x}_f^T F_f$$

$$= \frac{1}{2}(u_l^T u_l - v_l^T v_l - u_f^T u_f + v_f^T v_f)$$

$$= \frac{1}{2}(u_l^T u_l - u_l^T(t - T_1)u_l(t - T_1) + v_f^T v_f - v_f^T(t - T_2)v_f(t - T_2))$$

$$= \frac{d}{dt}(\frac{1}{2}\int_{t-T_1}^{t} u_l^T u_l d\sigma + \frac{1}{2}\int_{t-T_2}^{t} v_f^T v_f d\sigma).$$

$$(3.34)$$

The energy E stored in communication channels could be represented as follows:

$$E(t) = \int_0^t \dot{x}_l^T F_l - \dot{x}_f^T F_f d\sigma$$

$$= \frac{1}{2}(\int_{t-T_1}^{t} u_l^T u_l d\sigma + \int_{t-T_2}^{t} v_f^T v_f d\sigma).$$

$$(3.35)$$

When the time delays T_1 and T_2 are constant and the energy satisfies $E(t) > 0$, the system is passive [36], thus stability can be guaranteed. But when the time delay are varying, the proof for the system passivity is non-trivial and far more complicated [11], and the stability of the teleoperation system will be disturbed. Therefore, it is necessary to improve the traditional wave variable method.

(2) Wave variable method with time-varying delay

To handle the time-varying delay that exists in the communication channels of teleoperation systems, the wave variable correction-based method is adopted. Fig. 3.7 displays the wave variable correction-based method in a teleoperation system.

It also could be represented as follows:

$$\hat{U}_f(t) = U_l(t - T_1(t)) + \delta U_f(t), \qquad (3.36)$$

$$\hat{V}_l(t) = V_f(t - T_2(t)) + \delta V_l(t), \qquad (3.37)$$

where $\hat{U}_f$ and $\hat{V}_l$ are the compensated wave variable outputs. The corrective waves variables δU_f and δV_l are presented as [37]

$$\delta U_f = \sqrt{2b}\lambda[x_{lf}(t) + x_{dh} - x_{fd}(t)], \qquad (3.38)$$

$$\delta V_l = \sqrt{2b}\lambda[x_{ff}(t) + x_{dh} - x_{ld}(t)], \qquad (3.39)$$

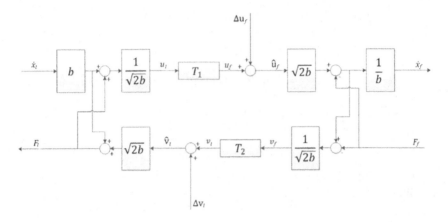

Figure 3.7 General wave variable method.

where $\lambda > 0$ guarantees convergence of the position and x_{dh} represents the initial position, which could be regarded as a zero vector when the gravity factor is not considered [37]. The desired positions x_{ld} and x_{fd} are described by

$$x_{ld} = \int_0^t \frac{(U_l(t) + \hat{V}_l(t))}{\sqrt{2b}} d\sigma, \qquad (3.40)$$

$$x_{fd} = \int_0^t \frac{(\hat{U}_f(t) + V_f(t))}{\sqrt{2b}} d\sigma. \qquad (3.41)$$

Thus, the desired position of the leader and follower is determined by the integrals of the paired waves $U_l - V_l$ and $U_f - V_f$, and its initial value is set as the current position of the robot. In addition, x_{lf} and x_{ff} are the fictitious positions of the robot, which could be represented as follows:

$$x_{lf}(t) = \frac{1}{\sqrt{2b}} \left(\int_0^{t-T_1(t)} U_l(t) d\sigma + \int_0^t V_f(t) d\sigma \right), \qquad (3.42)$$

$$x_{ff}(t) = \frac{1}{\sqrt{2b}} \left(\int_0^t U_l(t) d\sigma + \int_0^{t-T_2(t)} V_f(t) d\sigma \right). \qquad (3.43)$$

From Eqs. (3.38)–(3.39), it can be seen that the corrective waves are proportional to the difference between the desired position and the fictitious position, trying to converge the position difference between the trajectories to the ideal value, which will be particularly useful for trajectory and force control [37]:

$$x_{fd}(t) - x_{ld}(t - T_1(t)) \rightarrow \frac{1}{\sqrt{2b}} \int_{t-T_1(t)-T_2(t)}^{t} U_l(t)d\sigma \rightarrow 0, \qquad (3.44)$$

$$x_{ld}(t) - x_{fd}(t - T_2(t)) \rightarrow \frac{1}{\sqrt{2b}} \int_{t-T_1(t)-T_2(t)}^{t} U_l(t)d\sigma \rightarrow 0. \qquad (3.45)$$

3.4.2 Multiple channel-based wave variable method

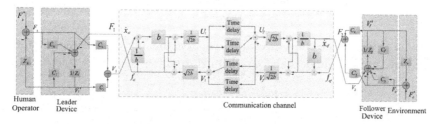

Figure 3.8 The proposed control architecture contains a four-channel scheme with the wave variable method for the teleoperation system.

Fig. 3.8 displays the proposed wave variable approach with a four-channel structure. Inspired by Refs. [38,39], the relationship of intermediate variables U_s and V_m can be represented as follows:

$$U_f(t) = 2U_d(t - T) - V_{cf}(t), \qquad (3.46)$$

$$V_l(t) = 2V_{cf}(t - T) + U_d(t). \qquad (3.47)$$

To reduce the impacts of wave reflections, impedance matching is employed. According to Fig. 3.8, U_l and V_f can be derived as follows:

$$\begin{aligned} U_l(t) &= \frac{1}{\sqrt{2b}}(f_{cl} + b(\dot{x}_{cl} - \frac{1}{b}f_{cl})) \\ &= \frac{b}{\sqrt{2b}}\dot{x}_{cl}, \end{aligned} \qquad (3.48)$$

$$\begin{aligned} V_f(t) &= \frac{1}{\sqrt{2b}}(b\dot{x}_{cf} - b(\dot{x}_{cf} + f_{cf})) \\ &= -\frac{1}{\sqrt{2b}}\dot{f}_{cf}. \end{aligned} \qquad (3.49)$$

Based on Fig. 3.8 and Eqs. (3.48)–(3.49), f_{cl}, V_l, $\dot{x}_{cf}$, and U_f can be represented as follows:

$$f_{cl} = \frac{1}{2}(b\dot{x}_{cl} - \sqrt{2b}\,V_l(t)), \tag{3.50}$$

$$V_l(t) = \frac{1}{\sqrt{2b}}(b\dot{x}_{cl} - 2f_{cl}), \tag{3.51}$$

$$\dot{x}_{cf} = \frac{1}{b}(\sqrt{2b}\,U_f(t) - (f_{cf} + b\dot{x}_{cf})), \tag{3.52}$$

$$U_f(t) = \frac{1}{\sqrt{2b}}(2b\dot{x}_{cf} + f_{cf}). \tag{3.53}$$

According to the above equations, we have the following expression for the wave variables under the four-channel structure:

$$\begin{cases} U_l(t) = \dfrac{b\dot{x}_{cl}}{\sqrt{2b}}, \\[2mm] V_f(t) = -\dfrac{f_{cf}}{\sqrt{2b}}, \\[2mm] V_l(t) = \dfrac{b\dot{x}_{cl} - 2f_{cl}}{\sqrt{2b}}, \\[2mm] U_f(t) = \dfrac{2b\dot{x}_{cf} + f_{cf}}{\sqrt{2b}}. \end{cases} \tag{3.54}$$

The wave variable method can be regarded as a two-port system. According to the passive theory, the power of the system can be calculated as follows:

$$\begin{aligned} P &= P_{in} + P_{out} \\ &= \dot{x}_{cl}^T f_{cl} - \dot{x}_{cf}^T f_{cf} \\ &\quad + U_l^T(t)U_l(t) + V_f^T(t)V_f(t) - U_l^T(t)V_l(t) + U_f^T(t)V_f(t). \end{aligned} \tag{3.55}$$

The energy E of the two-port system can be represented as

$$\begin{aligned} E &= \int_{t_0}^{\tau-t_0} (P_{in} + P_{out})\,dt \\ &= \int_{t_0}^{\tau-t_0} (\dot{x}_{cl}^T f_{cl} - \dot{x}_{cf}^T f_{cf})\,dt \\ &= \frac{1}{2}\left(\int_{t_0}^{\tau-t_0} (U_l^T U_l)\,dt + \int_{t_0}^{\tau-t_0} (V_f^T V_f)\,dt \right). \end{aligned} \tag{3.56}$$

When $E > 0$, the teleoperation system is passive, and therefore stability can be guaranteed. Based on the wave variable method [38], the four-channel architecture can be represented as

$$V_1 = C_3 F_h + C_1 V_l', \tag{3.57}$$

$$F_2 = C_2 F_e + C_4 V_f', \tag{3.58}$$

$$V_l' Z_l = -V_l' C_l + F_h(1 + C_6) - F_1, \tag{3.59}$$

$$V_f' Z_f = -V_f' C_f - F_e(1 + C_5) + V_2, \tag{3.60}$$

$$F_1 = F_h(1 + C_6) - V_l' Z_{tl}, \tag{3.61}$$

$$V_2 = F_e(1 + C_5) + V_f' Z_{te}. \tag{3.62}$$

According to the proposed four-channel control and wave variable method, we obtain

$$U_l(t) = \frac{b V_1}{\sqrt{2b}}, \tag{3.63}$$

$$V_f(t) = \frac{-F_2}{\sqrt{2b}}, \tag{3.64}$$

$$V_l(t) = \frac{b V_1 - 2 F_1}{\sqrt{2b}}, \tag{3.65}$$

$$U_f(t) = \frac{2b V_2 + F_2}{\sqrt{2b}}. \tag{3.66}$$

Based on Eq. (3.46) and Eqs. (3.63)–(3.66), we have

$$\frac{2b V_2(t) + F_2(t)}{\sqrt{2b}} = \frac{2b V_1(t - T)}{\sqrt{2b}} + \frac{F_2(t)}{\sqrt{2b}}, \tag{3.67}$$

$$-V_2(t) = -V_1(t - T),$$
$$-V_2 = -V_1 \exp^{-sT}. \tag{3.68}$$

According to Eq. (3.47) and Eqs. (3.63)–(3.66), we have

$$\frac{b V_1(t) - 2 F_1(t)}{\sqrt{2b}} = \frac{-2 F_2(t - T)}{\sqrt{2b}} + \frac{b V_1(t)}{\sqrt{2b}}, \tag{3.69}$$

$$F_1(t) = F_2(t - T),$$
$$F_1 = F_2 \exp^{-sT}. \tag{3.70}$$

Then, based on Eq. (3.70) and Eq. (3.68), we have

$$\begin{bmatrix} F_1 \\ -V_2 \end{bmatrix} = \begin{bmatrix} 0 & \exp^{-sT} \\ -\exp^{-sT} & 0 \end{bmatrix} \begin{bmatrix} V_1 \\ F_2 \end{bmatrix}. \tag{3.71}$$

Thus, the scattering norm can be derived as

$$H(s) = \begin{bmatrix} 0 & \exp^{-sT} \\ -\exp^{-sT} & 0 \end{bmatrix} = \begin{bmatrix} h_{11} & h_{12} \\ h_{21} & h_{22} \end{bmatrix}. \tag{3.72}$$

Let $s = j\omega$. Then

$$\overset{1}{\sup}^{\frac{1}{2}}[H^*(j\omega)H(j\omega)] = \overset{1}{\sup}^{\frac{1}{2}} \begin{bmatrix} 1 & 0 \\ 0 & 1 \end{bmatrix} = 1. \tag{3.73}$$

According to Refs. [40,41], $\sup^{\frac{1}{2}}[H^*(j\omega)H(j\omega)] = 1$ indicates that the teleoperation system is passive, thus the stability can be guaranteed.

3.5. Experimental case study

3.5.1 Experimental case with RBFNN

The experimental platform is set up with the Touch X haptic device as the leader device and a simulated Baxter robot arm as the follower robot (Figs. 3.9–3.10). The desired trajectory was generated by the stylus of the Touch X joystick and sent to the Baxter robot through the communication channels.

In Ref. [11], the controller for the follower robot is designed as a conventional PD controller, and position trajectory tracking was carried out. The human operator moved the stylus of the joystick from an initial position to the minimum value along the x-direction, then to the maximum value along the x-direction, and finally back to the initial position, for a time span of 0–3 s. Similar operations along the y-direction and the z-direction were performed for 3–6 s and 6–9 s. The trajectory tracking results for the follower robot with the conventional PD controller is shown in Fig. 3.11.

Then, the RBFNN compensation was added to the basic PD controller [11]. In the first 10 s, the NN was training. The weights for different joints converge to different values. The weights of Baxter arm joints S1, W0, and W2 (the three joints that are closest to the body) are close to

Figure 3.9 Experiment platform of the leader–follower teleoperation system.

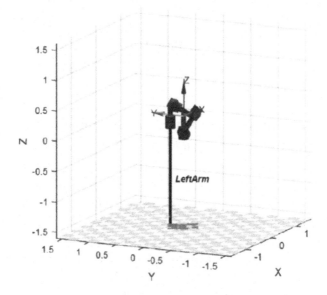

Figure 3.10 Simulated follower robot.

zero, for they are almost not affected by gravity during trajectory tracking. At 10–19 s, the human operator repeats the trajectory tracking process of the last experiment. The weights of the NN and the tracking performance with compensation are shown in Fig. 3.12 and Fig. 3.13.

The comparison of the above experiments shows that the tracking performance is much improved after adding the RBFNN compensation.

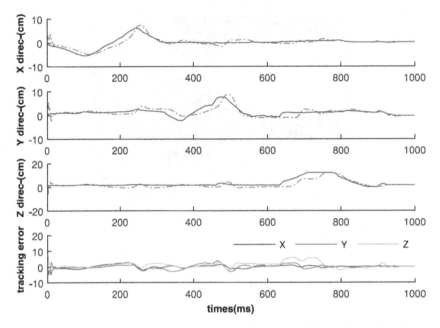

Figure 3.11 Trajectory of the leader device (solid) and the follower (dashed) with the conventional PD controller.

3.5.2 Experimental case with the wave variable method

(1) Wave variable correction-based method

With a time-varying delay in the communication channel, the teleoperation system is likely to become unstable and uncontrollable. In this section, we report experiments of trajectory tracking and force reflection under different communications to test the wave variable correction-based method and compare it with another experiment without using the wave variable [11]. The teleoperation system used in this experiment was consistent with the RBFNN case above.

The added time-varying delays T_1 and T_2 in this experiment are illustrated in Fig. 3.14. The force reflection to the human operator is generated by $F_{feedback} = K_{feedback}(x_f - x_l)$. The human operator was required to repeat the movement in the next two experiments. One group corresponded to the communication channel with wave variable compensation, and the other corresponded to no wave variable compensation. The trajectory tracking performance of the comparative experiment is shown in Fig. 3.15 and Fig. 3.16. We can find that in the trajectory tracking experiment, the group with the wave variable compensation shows less trajectory tracking

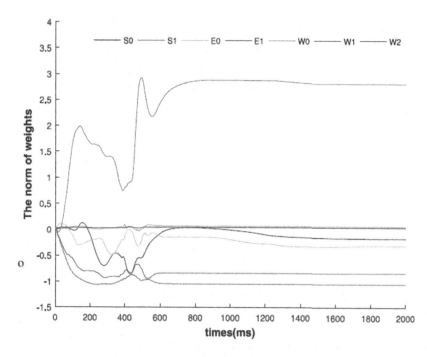

Figure 3.12 Norm of NN weights for each joint of the follower robot during training.

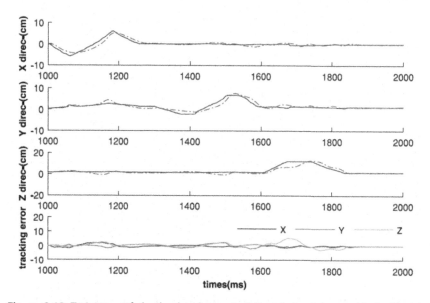

Figure 3.13 Trajectory of the leader device (solid) and the follower (dashed) with RBFNN compensation.

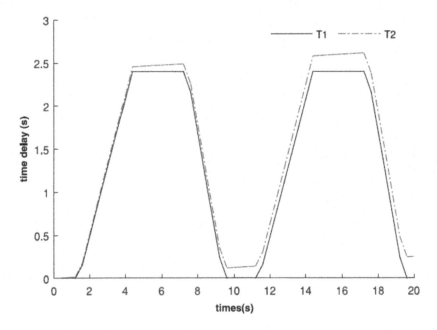

Figure 3.14 Varying time delays T_1 (solid) and T_2 (dashed).

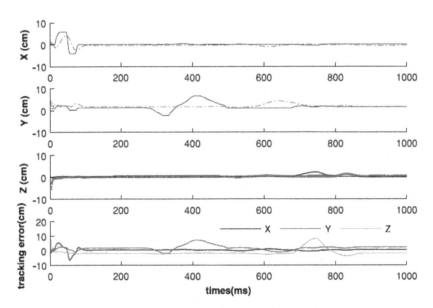

Figure 3.15 Trajectory of the leader device (solid) and the follower (dashed) with time-varying delayed communication, using the wave variable technique.

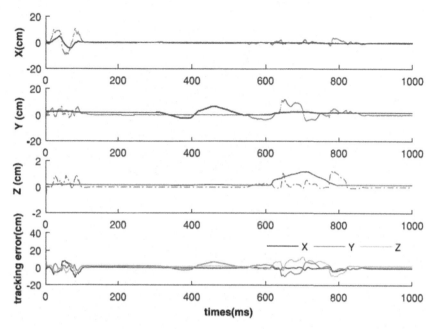

Figure 3.16 Trajectories of the leader device (solid) and the follower (dashed) with time-varying delayed communication, without using the wave variable technique.

error under the influence of varying time delays, which means the wave variable compensation is effective.

Then, in the force reflection experiment, a rigid workpiece similar to a wall was set up and installed along the x-direction (Fig. 3.17). In the first 3 s, both the leader device and the follower robot are in free motion. Then, the leader device starts to move to the position where the follower robot gets in touch with the workpiece, and the contact lasts for about 10 s. During the contact with the workpiece, the follower robot almost does not move, and the environmental force applied to the follower robot converges to a set value since the human operator held the leader device with a constant force. Then, the operator moved the leader joystick back to make the follower robot leave the workpiece and track the movement of the leader joystick without time-varying delay. Fig. 3.18 shows the trajectory and force reflection of the leader device and the follower robot in the x-direction. It can be seen that the system has a great force reflection effect during the time-varying delay, and the oscillation of the system is caused by the force reflection process before and after contact [11].

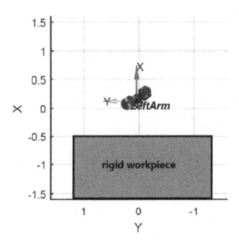

Figure 3.17 Simulated follower robot with workpiece.

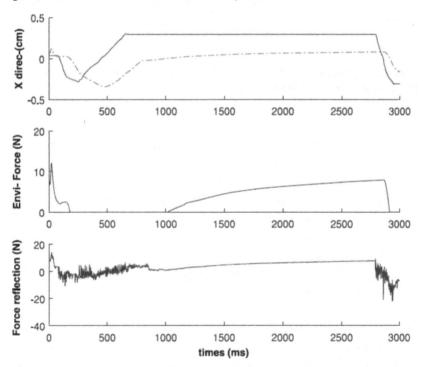

Figure 3.18 (Top) Trajectories along the x-direction of the leader device (solid) and of the follower (dashed) with time-varying delayed communication, using the wave variable technique. (Middle) Environmental force to the follower robot. (Bottom) Force reflections to the leader device along the x-direction with time-varying delayed communication, using the wave variable technique.

(2) Multiple channel-based wave variable method

The parameters of the teleoperation system are chosen based on a pilot experiment. Specifically, $b = 800$ N·s/m. The time delays are set as 200 ms, 400 ms, 800 ms, and 1200 ms. We assume that the external incentive F_h is a sine function. The follower robot follows the leader's movement. The parameters are selected as follows: $C_2 = C_3 = 0.495$, $C_5 = C_6 = -0.495$. The masses $M_l = M_f = 0.95$ kg, and $C_l = 50M_l(1+s)/s$, $C_f = 50M_f(1+s)/s$, $Z_l = M_l s$, $Z_f = M_f s$.

In order to verify the performance of the proposed method under different time delays, the mean absolute error (MAE) of the tracking performance is introduced. The MAE can be represented as

$$MAE = \frac{1}{N}\sum_{i=1}^{N}|z_i - \hat{z}_i|, \tag{3.74}$$

where z_i and $\hat{z}_i$ are the desired value and the actual value, respectively, and N represents the number of sample value.

Case 1 – Performance tracking under a time delay of 200 ms

Fig. 3.19(a) shows the position tracking performance of the teleoperation system with a time delay of 200 ms. Blue curves and red curves represent the performance of the leader and the follower, respectively. It can be seen that the curves are oscillating at the beginning, but the follower can completely track the position of the leader soon. In Fig. 3.19(b), the velocities of the system are very low, and the follower can track the leader perfectly since the system employs the same structure as the leader–follower framework.

Fig. 3.19(c) shows the force reflection performance of the system. It can be seen that the follower can track the leader perfectly. It can be concluded that the force line of the leader and the follower has nearly no distortion. The performance of force tracking demonstrates that transparency of the teleoperation system can be guaranteed.

Case 2 – Performance tracking under a time delay of 400 ms

The tracking performance of the leader and the follower with a time delay of 400 ms is presented in Fig. 3.20. Fig. 3.20(a and b) shows that the trajectory tracking performance of the follower tracks the leader effectively. However, the process of position tracking is unstable due to the relatively large initial position difference. A similar conclusion can be drawn: the force tracking of the follower can track the leader with no distortion in Fig. 3.20(c).

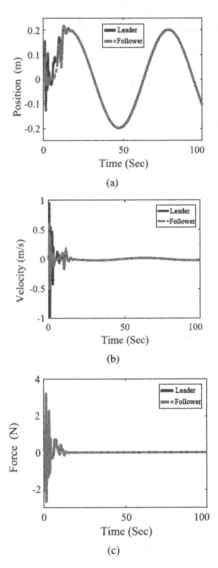

Figure 3.19 Tracking performance under a time delay of 200 ms. (a) Trajectory tracking for the leader (solid) and the follower (dash-dotted). (b) Velocity tracking for the leader and the follower. (c) Force tracking for the leader (solid) and the follower (dash-dotted).

Case 3 – Performance tracking under a time delay of 800 ms and 1200 ms

The tracking performance under time delays of 800 ms and 1200 ms is shown in Figs. 3.21(a)–3.22(c). In Fig. 3.21(a and b), the follower barely

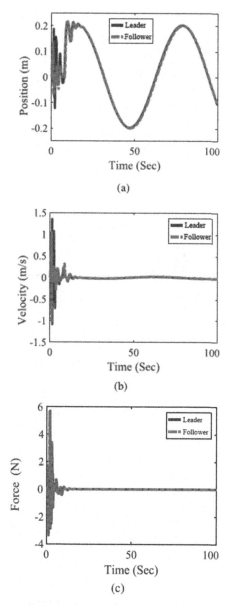

Figure 3.20 Tracking performance under a time delay of 400 ms. (a) Trajectory tracking for the leader (solid) and the follower (dash-dotted). (b) Velocity tracking for the leader and the follower. (c) Force tracking for the leader (solid) and the follower (dash-dotted).

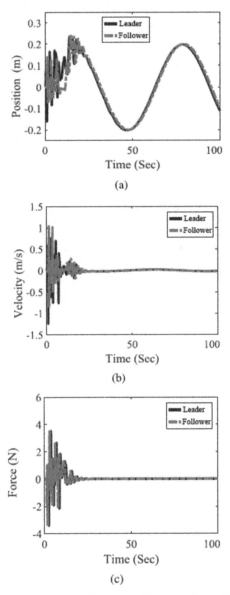

Figure 3.21 Tracking performance under a time delay of 800 ms. (a) Trajectory tracking for the leader (solid) and the follower (dash-dotted). (b) Velocity tracking for the leader and the follower. (c) Force tracking for the leader (solid) and the follower (dash-dotted).

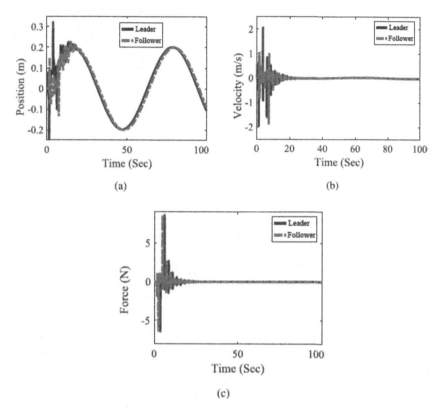

Figure 3.22 Tracking performance under a time delay of 1200 ms. (a) Trajectory tracking for the leader (solid) and the follower (dash-dotted). (b) Velocity tracking for the leader and the follower. (c) Force tracking for the leader (solid) and the follower (dash-dotted).

follows the movement of the leader. In Fig. 3.22(a and b), the system can still maintain stability, and there is an obvious delay in the tracking performance for the leader and the follower.

For the force tracking performance of the leader and the follower under time delays of 800 ms and 1200 ms, a similar conclusion can be drawn: the force tracking performance is distorted at the beginning and becomes stable towards the end (Figs. 3.21(c) and 3.22(c)).

Case 4 – Performance comparison under time delays of 200 ms, 400 ms, 800 ms, and 1200 ms

Compared with the tracking performance in Cases 1 and 2, the performance in Case 3 is much worse. For trajectory tracking, the performance under a time delay of 800 ms is better than that under a time delay of

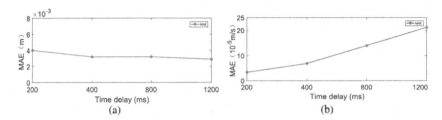

Figure 3.23 Performance analysis under different time delays. (a) MAE of position. (b) MAE of velocity.

Table 3.1 MAE of the tracking performance under different time delay conditions.

Time delay	Position	Velocity
200 ms	0.0040	3.3503×10^{-05}
400 ms	0.0032	6.8380×10^{-05}
800 ms	0.0032	13.916×10^{-05}
1200 ms	0.0029	21.090×10^{-05}

1200 ms. Table 3.1 and Fig. 3.23 show the MAE of position and velocity. It can be seen that the MAE values for position and velocity become larger with increasing time delays. We can conclude that the tracking performance deteriorates with increasing time delays. The tracking performance with a time delay of 200 ms is better than with time delays of 400 ms, 800 ms, and 1200 ms.

In this section, we perform a comparative test between a four-channel scheme and the proposed scheme. In this comparison, the delay time is set as 1000 ms. Figs. 3.24 and 3.25 show the tracking performance by using the four-channel scheme and the proposed method. It can be seen that the four-channel scheme and the proposed method can achieve good performance in the tracking test. However, the proposed method can enhance the position and velocity tracking with smaller MAE values than the four-channel scheme (Table 3.2).

Table 3.2 MAE of the tracking performance for the four-channel scheme and the proposed scheme.

Scheme	Position	Velocity
Four-channel	0.0204	0.0060
Proposed	0.0200	0.0042

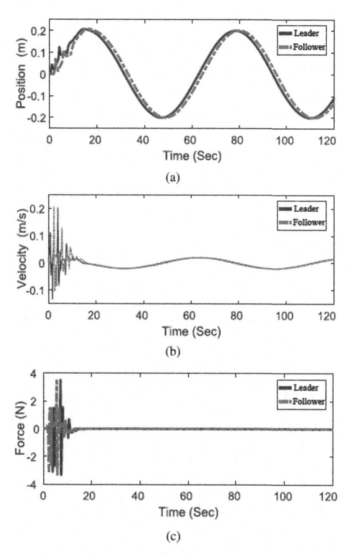

Figure 3.24 Tracking performance of the four-channel scheme. (a) Trajectory tracking for the leader (solid) and the follower (dash-dotted). (b) Velocity tracking for the leader and the follower. (c) Force tracking for the leader (solid) and the follower (dash-dotted).

3.6. Conclusion

In this chapter, we propose several methods based on wave variables and RBFNN to deal with the uncertainty problems in teleoperation sys-

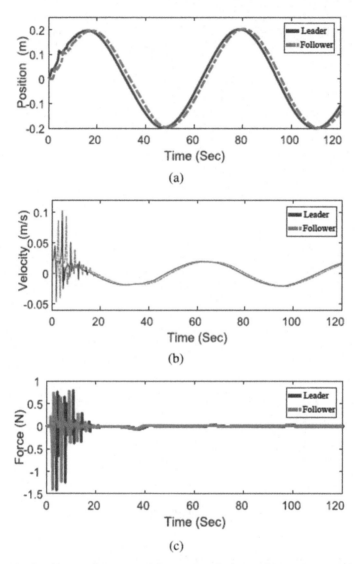

Figure 3.25 Tracking performance of the proposed scheme. (a) Trajectory tracking for the leader (solid) and the follower (dash-dotted). (b) Velocity tracking for the leader and the follower. (c) Force tracking for the leader (solid) and the follower (dash-dotted).

tems, which are mainly caused by the time delays in the communication channel and the dynamics in the actual environment.

For the nonlinearity of the dynamic model and the imprecision of the dynamic parameters, the RBFNN-based controller is proposed for compensation. With the advantages of nonlinear function approaching and fast training, RBFNN could be competent for approaching nonlinear items in the dynamic model and finally realize precise tracking. Experiments to determine tracking performance was carried out to analyze the effectiveness of RBFNN compensation.

For the time-varying delay existing in the communication channel, the wave variable correction-based method was proposed. The traditional wave variable method could ensure the passivity and stability of a system under a constant time delay, but when the delays are time-varying, the stability of the system is disturbed. Thus, the wave variable method with correction item was introduced in this chapter. Trajectory tracking and force reflection experiments showed that the correction-based wave variable method is stable and effective.

Besides, since the multichannel structure can improve the transparency and stability of bilateral teleoperation systems, it is necessary to consider the solution when the time delay exists in the communication channel under a multichannel structure. In this chapter, the existence of time delay in a four-channel structure was taken as an example, and we introduced a four-channel scheme-based wave variable method. Theoretical proofs were provided and experiments were carried out to show that the multiple-channel scheme-based wave variable method could ensure stability and transparency of the teleoperation system in the presence of time delay.

References

[1] Zhijun Li, Chun-Yi Su, Neural-adaptive control of single-master–multiple-slaves teleoperation for coordinated multiple mobile manipulators with time-varying communication delays and input uncertainties, IEEE Transactions on Neural Networks and Learning Systems 24 (9) (2013) 1400–1413.

[2] Zhenyu Lu, Panfeng Huang, Zhengxiong Liu, Relative impedance-based internal force control for bimanual robot teleoperation with varying time delay, IEEE Transactions on Industrial Electronics 67 (1) (2019) 778–789.

[3] F. Hong, S.S. Ge, B. Ren, T.H. Lee, Robust adaptive control for a class of uncertain strict-feedback nonlinear systems, International Journal of Robust and Nonlinear Control 19 (7) (2009) 746–767.

[4] Zhenyu Lu, Panfeng Huang, Zhengxiong Liu, Predictive approach for sensorless bimanual teleoperation under random time delays with adaptive fuzzy control, IEEE Transactions on Industrial Electronics 65 (3) (2017) 2439–2448.

[5] Zhenyu Lu, Panfeng Huang, Pei Dai, Zhengxiong Liu, Zhongjie Meng, Enhanced transparency dual-user shared control teleoperation architecture with multiple adaptive dominance factors, International Journal of Control, Automation, and Systems 15 (5) (2017) 2301–2312.

[6] Roberto Oboe, Paolo Fiorini, A design and control environment for internet-based telerobotics, The International Journal of Robotics Research 17 (4) (1998) 433–449.

[7] Nikhil Chopra, Paul Berestesky, Mark W. Spong, Bilateral teleoperation over unreliable communication networks, IEEE Transactions on Control Systems Technology 16 (2) (2008) 304–313.

[8] Hayata Sakai, Daisuke Tomizuka, Kouhei Ohnishi, Compliance control for stabilization of bilateral teleoperation system in the presence of time delay, in: 2017 IEEE International Conference on Mechatronics (ICM), IEEE, 2017, pp. 62–67.

[9] Neal A. Tanner, Günter Niemeyer, Improving perception in time-delayed telerobotics, The International Journal of Robotics Research 24 (8) (2005) 631–644.

[10] Panfeng Huang, Pei Dai, Zhenyu Lu, Zhengxiong Liu, Asymmetric wave variable compensation method in dual-master-dual-slave multilateral teleoperation system, Mechatronics 49 (2018) 1–10.

[11] Chenguang Yang, Xingjian Wang, Zhijun Li, Yanan Li, Chun-Yi Su, Teleoperation control based on combination of wave variable and neural networks, IEEE Transactions on Systems, Man, and Cybernetics: Systems 47 (8) (2016) 2125–2136.

[12] Yuan Yuan, Yingjie Wang, Lei Guo, Force reflecting control for bilateral teleoperation system under time-varying delays, IEEE Transactions on Industrial Informatics 15 (2) (2018) 1162–1172.

[13] Servet Soyguder, Tayfun Abut, Haptic industrial robot control with variable time delayed bilateral teleoperation, Industrial Robot (2016).

[14] Yongqiang Ye, Peter X. Liu, Improving haptic feedback fidelity in wave-variable-based teleoperation orientated to telemedical applications, IEEE Transactions on Instrumentation and Measurement 58 (8) (2009) 2847–2855.

[15] Yongqiang Ye, Peter X. Liu, Improving trajectory tracking in wave-variable-based teleoperation, IEEE/ASME Transactions on Mechatronics 15 (2) (2009) 321–326.

[16] Dale A. Lawrence, Stability and transparency in bilateral teleoperation, IEEE Transactions on Robotics and Automation 9 (5) (1993) 624–637.

[17] Da Sun, Fazel Naghdy, Haiping Du, Transparent four-channel bilateral control architecture using modified wave variable controllers under time delays, Robotica 34 (4) (2016) 859–875.

[18] Da Sun, Fazel Naghdy, Haiping Du, Wave-variable-based passivity control of four-channel nonlinear bilateral teleoperation system under time delays, IEEE/ASME Transactions on Mechatronics 21 (1) (2015) 238–253.

[19] Amir Haddadi, Keyvan Hashtrudi-Zaad, Robust stability of teleoperation systems with time delay: A new approach, IEEE Transactions on Haptics 6 (2) (2012) 229–241.

[20] Phongsaen Pitakwatchara, Wave correction scheme for task space control of time-varying delayed teleoperation systems, IEEE Transactions on Control Systems Technology 26 (6) (2017) 2223–2231.

[21] Jing Luo, Chao Liu, Ning Wang, Chenguang Yang, A wave variable approach with multiple channel architecture for teleoperated system, IEEE Access 7 (2019) 143912–143920.

[22] Yan-Jun Liu, Shaocheng Tong, Adaptive fuzzy control for a class of nonlinear discrete-time systems with backlash, IEEE Transactions on Fuzzy Systems 22 (5) (2014) 1359–1365.

[23] Di-Hua Zhai, Yuanqing Xia, Adaptive control for teleoperation system with varying time delays and input saturation constraints, IEEE Transactions on Industrial Electronics 63 (11) (2016) 6921–6929.

[24] Long Cheng, Zeng-Guang Hou, Min Tan, Yingzi Lin, Wenjun Zhang, Neural-network-based adaptive leader-following control for multiagent systems with uncertainties, IEEE Transactions on Neural Networks 21 (8) (2010) 1351–1358.

[25] Chenguang Yang, Yiming Jiang, Zhijun Li, Wei He, Chun-Yi Su, Neural control of bimanual robots with guaranteed global stability and motion precision, IEEE Transactions on Industrial Informatics 13 (3) (2016) 1162–1171.

[26] Wei He, Zhao Yin, Changyin Sun, Adaptive neural network control of a marine vessel with constraints using the asymmetric barrier Lyapunov function, IEEE Transactions on Cybernetics 47 (7) (2016) 1641–1651.

[27] Yu Lu, J.K. Liu, F.C. Sun, Actuator nonlinearities compensation using RBF neural networks in robot control system, in: Proceedings of the Multiconference on "Computational Engineering in Systems Applications", vol. 1, IEEE, 2006, pp. 231–238.

[28] Jun Li, Tom Duckett, Some practical aspects on incremental training of RBF network for robot behavior learning, in: 2008 7th World Congress on Intelligent Control and Automation, IEEE, 2008, pp. 2001–2006.

[29] Chenguang Yang, Xingjian Wang, Zhijun Li, Yanan Li, Chun-Yi Su, Teleoperation control based on combination of wave variable and neural networks, IEEE Transactions on Systems, Man, and Cybernetics: Systems 47 (8) (2017) 2125–2136.

[30] Da Sun, Fazel Naghdy, Haiping Du, Neural network-based passivity control of teleoperation system under time-varying delays, IEEE Transactions on Cybernetics 47 (7) (2017) 1666–1680.

[31] Lorenzo Sciavicco, Bruno Siciliano, A dynamic solution to the inverse kinematic problem for redundant manipulators, in: Proceedings of the 1987 IEEE International Conference on Robotics and Automation, vol. 4, IEEE, 1987, pp. 1081–1087.

[32] Yusheng T. Tsai, David E. Orin, A strictly convergent real-time solution for inverse kinematics of robot manipulators, Journal of Robotic Systems 4 (4) (1987) 477–501.

[33] Lorenzo Sciavicco, Bruno Siciliano, Solving the inverse kinematic problem for robotic manipulators, in: RoManSy 6, Springer, 1987, pp. 107–114.

[34] Jean-Jacques Slotine, Dana Yoerger, A rule-based inverse kinematics algorithm for redundant manipulators, International Journal of Robotics and Automation 2 (2) (1987) 86–89.

[35] Lorenzo Sciavicco, Bruno Siciliano, A solution algorithm to the inverse kinematic problem for redundant manipulators, IEEE Journal of Robotics and Automation 4 (4) (1988) 403–410.

[36] Jean-Jacques E. Slotine, Weiping Li, et al., Applied Nonlinear Control, vol. 199, Prentice Hall, Englewood Cliffs, NJ, 1991.

[37] Phongsaen Pitakwatchara, Control of time-varying delayed teleoperation system using corrective wave variables, in: 2015 IEEE/RSJ International Conference on Intelligent Robots and Systems (IROS), IEEE, 2015, pp. 4550–4555.

[38] Hongbing Li, Kenji Kawashima, Achieving stable tracking in wave-variable-based teleoperation, IEEE/ASME Transactions on Mechatronics 19 (5) (2013) 1574–1582.

[39] Jing Guo, Chao Liu, Philippe Poignet, Scaled position-force tracking for wireless teleoperation of miniaturized surgical robotic system, in: 2014 36th Annual International Conference of the IEEE Engineering in Medicine and Biology Society, IEEE, 2014, pp. 361–365.

[40] Robert J. Anderson, Mark W. Spong, Bilateral control of teleoperators with time delay, in: Proceedings of the 1988 IEEE International Conference on Systems, Man, and Cybernetics, vol. 1, IEEE, 1988, pp. 131–138.

[41] Zheng Chen, Fanghao Huang, Weichao Sun, Wei Song, An improved wave-variable based four-channel control design in bilateral teleoperation system for time-delay compensation, IEEE Access 6 (2018) 12848–12857.

CHAPTER FOUR

User experience-enhanced teleoperation control

In this chapter, we will give a brief review on the state-of-the-art developments in electromyographic (EMG) stiffness transfer from humans to robots. Subsequently, in Section 4.2, we will introduce several commonly used approaches that can model human skills and thus enable skill transfer to robots. In Section 4.3, the design of some intelligent controllers that can enhance robotic skill learning will be introduced. In Section 4.4, we will provide a summary of this chapter.

4.1. Introduction

In order to improve the reliability and controllability of teleoperation robot systems, scholars have proposed many control strategies in recent years, for example, the H-infinity controller which realizes hybrid position and force control [1] and the new leader–follower controller based on Takagi–Sugeno fuzzy logic in order to ensure stability of the system [2]. Other control methods such as inversion control [3] and neural network (NN) adaptive control [4] are also used to improve the stability, robustness, and accuracy of teleoperation robot systems. On one hand, since the traditional control method is mainly designed for control of the robot body, the operator is not taken into consideration. On the other hand, these control strategies only involve the performance improvement between a single robot and the operator, and there is a lack of timely two-way information feedback between the operator and the robot. Therefore, the human–robot interactive control of teleoperation robots is still a challenge. For the above problems, the methods described in this chapter use EMG signal changes during the operation of the operator as an "indicator" to feed back the performance of muscle activity to the teleoperation robot control system so that the robot can "perceive" the operator's intentions and natural, friendly human–robot interaction (HRI) control can be achieved.

Physiological tremor, as a kind of involuntary movement, exists in every part of the human body, especially the limbs and head. Physiological tremor

73

has no effect on people's daily life, but in scenes that require flexible operation, tremor will inevitably affect the performance of teleoperation [5]. Some researchers have proposed methods based on autoregressive processes [6] and a band-limited multiple Fourier linear combiner [7] to deal with the tremor problem. However, the autoregression method regards the tremor signal as a linear Gaussian random process, which weakens the characteristics of the tremor signal. It is difficult for a band-limited multiple Fourier linear combiner to determine its own network structure and optimal solution. In addition, the autoregression method and the band-limited multiple Fourier linear combiner need to collect a large amount of sample data to build a model of the tremor signal, which is difficult to implement in practice. To solve this problem, in this chapter we will use support vector machines (SVMs) to solve the problem of small training data samples and construct a tremor filter.

It is difficult to provide sufficient real-time perception for teleoperation systems. Moreover, due to the unskillful operation and muscle physiological tremor of the human operator, the natural performance cannot guarantee secure operation. Thus, it is important to enhance the interaction capability of the teleoperation system. Virtual fixture (VF) is an alternative method to improve teleoperation performance. Virtual fixture was proposed to extract relevant information between the human operator and the remote environment for HRI [8]. In Ref. [9], a VF control strategy was presented to improve the manipulation performance of an active handrest. Becker, MacLachlan, Lobes, Hager, and Riviere developed a derivation of VF based on the motion of the instrument in real-time for system control [10]. In Ref. [11], the authors proposed a forbidden region VF with a robust fuzzy logic controller to improve the human manipulation performance during laparoscopic surgery. A VF method based on the position error was presented to add an augmentation force on the leader device to improve the task quality [12]. Selvaggio et al. proposed an online VF and task switching mechanism that utilizes a stereo camera system to provide position information [13], thus improving teleoperation performance. In Ref. [14], a flexible VF method with force vision-based scheme was developed to reduce cognitive load and improve the task performance. In this chapter, due to unskillful operation and muscle physiological tremor of the human operator, a VF method is developed to ensure accuracy of operation and to reduce the operation pressure on the human operator.

In practical applications, the robotic manipulator is often subjected to uncertain external disturbance. When perturbation is applied on the robot, large errors will appear in the tracking task. To reduce the effect of disturbance, researchers put forward many methods for different kinds of disturbance in various systems. A hybrid scheme with an iterative learning composite anti-disturbance structure was proposed to handle model uncertainties and link vibration of manipulators in Ref. [15]. A distributed adaptive fuzzy algorithm was proposed to cope with nonlinear system uncertainties in Ref. [16]. A cerebellar model articulation controller was designed to overcome the effects of parameter uncertainties and external disturbances in free-floating space manipulators (FFSMs) [17]. In Ref. [18], time delay and its effects leading to system uncertainties or disturbance were considered, and in Ref. [19], a linear filter to estimate the admissible parameter uncertainties was designed. Among these techniques in the scholar society, the disturbance observer (DOB) is an efficient and widely used technology to improve the tracking performance. The DOB technology was proposed by Ohnishi et al. [20], and due to its robustness, it can be intuitively adjusted in a desired bandwidth, which plays an important role in robust control [21]. Moreover, since it is effective in compensating for the influence of disturbance and model uncertainties, it has been widely used in robotics, industrial automation, and the automotive industry [22,23]. In Ref. [24], DOB was applied in a teleoperation system by attenuating the influence of disturbances. DOB is able to compensate for the model uncertainties by estimating external disturbances [25,26]. A novel nonlinear DOB (NDOB) for robotic manipulation was proposed in Ref. [27], and the effectiveness of NDOB was demonstrated by controlling a two-link robotic arm. A three-link robotic manipulator was controlled by NDOB in Ref. [28], which was the extension of [27], and the stability of NDOB was verified by the direct Lyapunov method. In Ref. [29], a scheme with adaptive fuzzy tracking ability was designed for multiinput and multioutput (MIMO) nonlinear systems under the circumstances with unknown non-symmetric input saturation, system uncertainties, and external disturbances. DOB has the advantage of simplicity and the ability to compensate for model uncertainties. However, when the dynamic model is coupled with fast time-varying perturbations, it is not adequate to compensate for all uncertainties or disturbances.

To tackle the problems concerning robot dynamics uncertainties, the human limb stiffness transfer schemes and adaptive control techniques have been utilized in Refs. [30,31]. However, human limb stiffness transfer and adaptive control might not do well in transient performance or cannot address the influence resulting from unparameterizable external disturbances through a parameter adaptation algorithm. To overcome high-frequency perturbations, in Ref. [32], the authors proposed a control strategy that integrates robot automatic control and human–operated impedance control using stiffness by a DOB-based adaptive control technique. However, this method does not take advantage of the motion skills of the human, and the movement of the telerobot cannot be controlled. In Ref. [33], to control the variable stiffness joints robot, a DOB-based adaptive NN control was proposed. Thereby, to enhance tracking performance in a teleoperation system, traditional DOB is insufficient. In this chapter, for the purpose of constructing a feasible teleoperation scheme, we develop a novel controller that contains a variable gain scheme to deal with fast time-varying perturbation, whose gain is linearly adjusted according to human surface EMG (sEMG) signals collected from MYO Armbands.

4.2. Variable stiffness control and tremor attenuation

4.2.1 Description of the teleoperation robot system

A teleoperation robot system is mainly composed of a leader part and a follower part. The leader part includes the leader computer, haptic device, and MYO Armband, and the follower part is composed of the follower computer and the follower device. The overall framework of the teleoperation robot system is shown in Fig. 4.1.

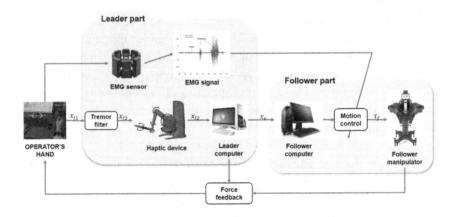

Figure 4.1 Block diagram of the teleoperation robot system.

4.2.2 Problem description

Undoubtedly, the physiological tremor of the operator greatly affects the stability and precision of a teleoperation system. Besides, the degrees of freedom (DOFs) of the Touch X device may influence the telerobot. To clarify the connection between the Touch X device (six DOFs) and the tremor, the modified Denavit–Hartenberg (D-H) method in Ref. [34] will be embraced in this section. In the presence of tremor, the uniform transformation matrix between the ith joint and the $(i-1)$th joint of the Touch X device can be expressed as follows:

$$\begin{cases} \theta_i + \Delta\theta_i = \tilde{\theta}_i, \ i = 1, 2, 3, ..., 6, \\ d_i + \Delta d_i = \tilde{d}_i, \ i = 1, 2, 3, ..., 6, \\ s = sin, \\ c = cos, \end{cases} \tag{4.1}$$

where θ_i and d_i are the expected value of each joint angle and the expected distance, respectively, and $\Delta\theta_i$ and Δd_i are the disturbances caused by the tremor.

Because of the influence of the physiological tremor, the reliability and stability of the motion of the end-effector may not be guaranteed. The six matrices ${}^0_1\tilde{T}, {}^2_1\tilde{T}, {}^3_2\tilde{T}, {}^4_3\tilde{T}, {}^5_4\tilde{T}$, and ${}^6_5\tilde{T}$ of the end-effector can be multiplied in a sequence as follows:

$$ {}^0_6\tilde{T} = {}^1_0\tilde{T}\,{}^2_1\tilde{T}\,{}^3_2\tilde{T}\,{}^4_3\tilde{T}\,{}^5_4\tilde{T}\,{}^6_5\tilde{T} = \begin{bmatrix} \tilde{n}_{11} & \tilde{n}_{12} & \tilde{n}_{13} & \tilde{p}_x \\ \tilde{n}_{21} & \tilde{n}_{22} & \tilde{n}_{23} & \tilde{p}_y \\ \tilde{n}_{31} & \tilde{n}_{32} & \tilde{n}_{33} & \tilde{p}_z \\ 0 & 0 & 0 & 1 \end{bmatrix}, \tag{4.2}$$

where $\tilde{n}_{11}, \tilde{n}_{12}, \tilde{n}_{13}, \tilde{n}_{21}, \tilde{n}_{22}, \tilde{n}_{23}, \tilde{n}_{31}, \tilde{n}_{32}$, and $\tilde{n}_{33}$ are the rotational elements of ${}^0_6\tilde{T}$ and $\tilde{p}_x, \tilde{p}_y$, and $\tilde{p}_x$ are the position vector of the pose matrix of the end effector ${}^0_6\tilde{T}$. We have

$$ {}^i_{i-1}\tilde{T} = \begin{bmatrix} c\tilde{\theta}_i & -s\tilde{\theta}_i & 0 & a_{i-1} \\ s\tilde{\theta}_i c\alpha_{i-1} & c\tilde{\theta}_i c\alpha_{i-1} & -s\alpha_{i-1} & -s\alpha_{i-1}\tilde{d}_i \\ s\tilde{\theta}_i s\alpha_{i-1} & c\tilde{\theta}_i s\alpha_{i-1} & c\alpha_{i-1} & c\alpha_{i-1}\tilde{d}_i \\ 0 & 0 & 0 & 1 \end{bmatrix}, \ i = 1, ..., 6. \tag{4.3}$$

4.2.3 Design and analysis for teleoperation

(1) Control strategy

Generally, each operator may have distinct muscle activations. To accomplish individualized control performance, the sEMG signals of the operator are adopted to estimate muscle activation in this section. With the muscle activation, the control gain changes as the individual operator. Taking into account the connection between control gain and muscle activation, the control gain is changed "individually" by different operators.

(1) Control stiffness calculation. With the muscle activation changing, the control stiffness changes accordingly. Because of the influence of the control strategy and the human stiffness, it is significant to normalize the control stiffness within a reasonable range. To accomplish the personalized control performance, the control stiffness of the ith sampling moment can be presented as

$$G_i = (G_{max} - G_{min})\frac{(a_i - a_{min})}{(a_{max} - a_{min})} + G_{min}, \qquad (4.4)$$

where G_{max} and G_{min} are the maximum and minimum stiffness that can be chosen guaranteeing stable robot motion, according to the designer's experience, and a_{max} and a_{min} are the maximum and minimum muscle activation, respectively. The values of G_{max}, G_{min}, a_{max}, and a_{min} can be acquired by pilot experiments.

(2) PD control design. PD controllers are adopted as both the controller for the leader and that for the follower. The PD controller of the teleoperation robot can be expressed as follows:

$$\begin{cases} \tau_l = K_l(q_l - q_{ld}) + D_l(\dot{q}_l - \dot{q}_{ld}), \\ \tau_f = K_f(q_f - q_{fd}) + D_f(\dot{q}_f - \dot{q}_{fd}), \end{cases} \qquad (4.5)$$

where τ_l and τ_f are the leader torque and the follower torque, respectively, K_l, D_l, K_f, and D_f are the PD controller parameters, and q_l and q_f are the actual joint angles for the leader and the follower, respectively. Accordingly, q_{ld} and q_{fd} are the desired joint angles.

(3) Haptic feedback. Haptic feedback varies with position error in the process of teleoperation, when the follower follows the movement of the leader. With increasing position error, the strength of the haptic feedback grows accordingly.

The generation of haptic feedback can be expressed as follows:

$$F_h = K_f(x_l - x_f), \qquad (4.6)$$

where K_f is the scaling factor of the haptic feedback and x_l and x_f are the leader device position and the follower device position, respectively.

(2) Design of the SVM filter

Compared with traditional filters, the proposed SVM filter introduces the structural risk minimization (SRM) principle, which enables the filter to solve the problem of small-scale sample data more effectively. In addition, the SVM filter could work without the parameters of the filter.

(1) Data processing. As an involuntary time-varying signal with low frequency, physiological tremor has a smooth curve and a relatively stable changing trend. When sampling signals, the contributions of sample sets are diverse in different time periods $\{x_i, y_i\}$. Thus, according to the properties of samples, we could design different weights m_k for the sample sets $\{x_i, y_i\}$ for different sampling times.

The linear weight function can be expressed as follows:

$$m_k = \frac{1 - d_k}{t_n - t_1} t_k + \frac{t_n d_k - t_1}{t_n - t_1}, \quad k = 1, 2, ..., N, \qquad (4.7)$$

where m_k is a linear function for the SVM model, $d_k \in (0, 1)$ is the contribution degree of the sample which needs to be tuned according to the tremor, and $t_k \in [0, +\infty)$ is the variable that indicates the time for the weight function.

(2) Structure of the SVM filter. In the whole filter system, the IMU module of the sampling model analyzes the actual trajectory instructed by the operator. An attenuation model is designed to attenuate the hand tremor, including its frequency, amplitude, and phase. The control model transforms the inverse kinematics into motion control variables for the leader robot. The mathematical model of the SVM filter is shown in Fig. 4.2.

In Fig. 4.2, $u(k)$ is the desired EMG signal and $n(k)$ and $\hat{n}(k)$ are the tremor signals and estimation values of $n(k)$, respectively. We have

$$S(k) = u(k) + n(k), \qquad (4.8)$$

where $(S_k, n_k)_k^N$, $k = 1, 2, ..., N$, are the inputs of the SVM model, and we can obtain $S(k) = S_k$, $n(k) = n_k$. The output of the SVM model $y(k)$ can be expressed as

$$y(k) = S(k) - \hat{n}(k) = u(k) + n(k) - \hat{n}(k). \qquad (4.9)$$

From the above equations, we can obtain $y(k) = u(k)$ if the tremor signals can be canceled completely. However, it is difficult to completely eliminate the tremor signals in practice.

(3) The optimization criterion of SVM. Based on the SVM model, a linear approximation map can be used to estimate the tremor signal [35]:

$$\hat{n}(k) = \omega^T \phi(S) + b, \qquad (4.10)$$

where $\omega \in R^n$ and $b \in R$ are the weight parameters and the bias term, respectively, and $\phi(\cdot)$ is a mapping that maps the original training data from a low-dimensional space to a high-dimensional feature space. For the given training data sets (S_i, n_i) and any $\varepsilon > 0$, there exists a hyperplane $n(k) = \omega^T \phi(x) + b$ that satisfies

$$|n(k) - \hat{n}(k)| \leq \varepsilon. \qquad (4.11)$$

Thus, the original problem can be transformed into an optimization problem as follows:

$$min \quad \frac{1}{2}\omega^T\omega + \frac{1}{2}Cm_k \sum_{i=1}^{N} \xi_i^2, \qquad (4.12)$$

$$subject\ to \quad n_i = \omega^T\phi(S) + b + \xi_i,$$

where n_i is the potential proximate linear mapping, m_k is the weight which is related to the tremor, and ξ_i and C are a positive slack variable and a regularization constant, respectively.

(4) Implementation of the SVM algorithm. Referring to Eq. (4.12), the optimization problem is described as follows:

$$L(\omega, b, \xi, \lambda) = \frac{1}{2}\omega^T\omega + \frac{1}{2}Cm_k \sum_{i=1}^{N} \xi_i^2$$

$$- \sum_{i=1}^{N} \lambda_i [\omega^T\phi(S_i) + b + \xi_i - n_i], \qquad (4.13)$$

where $\lambda_i \in R$ is the Lagrangian multiplier.

Differentiating both sides, $L(\omega, b, \xi, \lambda)$ must satisfy $\frac{\partial L}{\partial \omega} = 0$, $\frac{\partial L}{\partial b} = 0$, $\frac{\partial L}{\partial \xi_i} = 0$, and $\frac{\partial L}{\partial \lambda_i} = 0$, and we have

$$
\begin{cases}
\omega = \sum_{i=1}^{N} \lambda_i \phi(S_i), \\
\sum_{i=1}^{N} \lambda_i = 0, \\
\lambda_i = C m_k \xi_i, \quad i = 1, 2, ..., N, \\
\omega^T \phi(S_i) + b + \xi_i - n_i = 0.
\end{cases}
\tag{4.14}
$$

By solving the equation and eliminating ω and ξ_i, we obtain

$$
(\sum_{i=1}^{N} \lambda_i \phi(S_i)^T \phi(S_j) + b + \frac{\lambda_i}{C m_k} = n_i, \quad i, j = 1, 2, ..., N.
\tag{4.15}
$$

The above equations can be expressed in matrix form as follows:

$$
\begin{bmatrix} 0 & \vec{1}^T \\ \vec{1} & Q + (C m_k)^{-1} E \end{bmatrix} \begin{bmatrix} b \\ \lambda \end{bmatrix} = \begin{bmatrix} 0 \\ n \end{bmatrix},
\tag{4.16}
$$

where E is an identity matrix and

$$
\begin{cases}
\vec{1} = [1, 1, ..., 1]^T, \\
Q = \phi(S_i)^T \phi(S_j), \\
\lambda = [\lambda_1, \lambda_2, ..., \lambda_N]^T, \\
n = [n_1, n_2, ..., n_N]^T.
\end{cases}
\tag{4.17}
$$

From Eq. (4.16), one has

$$
\begin{aligned}
\phi &= \begin{bmatrix} 0 & \vec{1}^T \\ \vec{1} & Q + (C m_k)^{-1} E \end{bmatrix} \\
&= \begin{bmatrix} 0 & 1 & \cdots & 1 \\ 1 & K(S_1, S_1) + (C m_k)^{-1} & \cdots & K(S_1, S_N) \\ \cdots & \cdots & \cdots & \cdots \\ 1 & K(S_N, S_1) & \cdots & K(S_N, S_N) + (C m_k)^{-1} \end{bmatrix}.
\end{aligned}
\tag{4.18}
$$

Therefore, we obtain

$$
\begin{bmatrix} b \\ \lambda \end{bmatrix} = (Q)^{-1} \phi^T \begin{bmatrix} 0 \\ n \end{bmatrix},
\tag{4.19}
$$

$$Q_{i,j} = \phi(S_i)^T \phi(S_j) = K(S_i, S_j), \quad i,j = 1, 2, ..., N, \qquad (4.20)$$

where $K(S_i, S_j)$ is a kernel function of the SVM.

In this section, a hybrid kernel function is introduced to improve the performance of the proposed SVM filter:

$$
\begin{aligned}
K(S, S_j)_{Hybrid} &= K(S, S_j)_{RBF} + K(S, S_j)_{Poly} \\
&= \alpha exp(-\frac{|S - S_j|^2}{\sigma^2}) + (1 - \alpha)[S \cdot S_j + 1]^q,
\end{aligned}
\qquad (4.21)
$$

where $\sigma^2 > 0$ and $q > 0$ are the kernel width factor of the radial basis function (RBF) and the order of the polynomial, respectively, and $\alpha \in [0, 1]$ is the coefficient of the hybrid kernel function. In the hybrid kernel function $K(S, S_j)_{Hybrid}$, σ, q, and α need to be tuned.

Lemma 1. *For any dual function $K(x, y)$, if it is an inner product computation of a feature space, $K(x, y)$ must satisfy the following condition. For any $g(x) \neq 0$ and $\int g^2(x)dx < \infty$, we can obtain*

$$\iint K(x, y)g(x)g(x')dxdx' \geq 0. \qquad (4.22)$$

According to Lemma 1, the hybrid kernel function can meet the Mercer condition.

As shown in Fig. 4.2, we can obtain the model of the SVM filter, and its output can be expressed as follows:

$$n(S) = \sum_{i=1}^{N} \lambda_i K(S, S_i) + b. \qquad (4.23)$$

The entire filtering process of the SVM filter is shown in Fig. 4.2.

4.3. Hybrid control

4.3.1 Control scheme

A hybrid position/stiffness control method is introduced in this section.

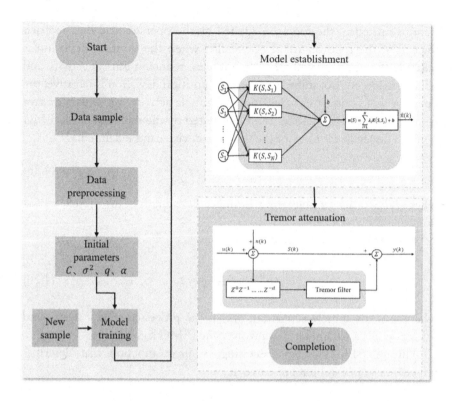

Figure 4.2 The architectures for SVM learning.

(1) PD control

A PD controller can be represented as

$$
\begin{aligned}
F_{pd} &= F^p + F^d \\
&= K_p(x_d - x_f) - K_d J_f(q_f)\dot{q}_f \\
&= K_p \tilde{x}_e - K_d J_f(q_f)\dot{q}_f,
\end{aligned}
\tag{4.24}
$$

where K_p and K_d are positive parameters of the PD controller, $\tilde{x}_e = x_d - x_f$ is the deviation between the desired trajectory x_d and the actual trajectory x_f, and $J_f(q_f)$ and q_f are the Jacobian matrix and joint displacement vector of the follower device, respectively.

(2) Variable stiffness control based on muscle activation

According to the external force applied in the follower manipulator, the operator can automatically adjust the muscle activation, whose changing trend is related to the force F_H applied by the operator and the feedback force f of the follower robot. Generally, when the position deviation is positive, the operator will relax the muscles, reducing muscle activity and obtaining less control stiffness. When the position deviation is negative, the operator will contract muscles, increase muscle activity, and obtain greater control stiffness. Both modes can be reflected by the intensity of the EMG signal. The feedback force of the follower device can be defined as

$$F^\alpha = K^\alpha (x_d - x_f), \tag{4.25}$$

where $K^\alpha > 0$ is a variable gain that indicates strength of the muscle activation.

The sEMG signal can be described as

$$\alpha^k = \frac{e^{\beta u(k)} - 1}{e^\beta - 1}, \tag{4.26}$$

where α is the muscle activation, $u(k)$ is the processed sEMG signal, and $-3 < \beta < 0$ is a parameter that involves the sEMG signal.

Through sEMG signal processing, a linear function that describes impedance gain can be represented as

$$K^\alpha = (K^\alpha_{max} - K^\alpha_{min}) \frac{(\alpha^k_i - \alpha^k_{max})}{(\alpha^k_{min} - \alpha^k_{max})} + K^\alpha_{min}. \tag{4.27}$$

(3) Integrated control

In the human–robot interaction, the control force F^r involves the muscle activation F^a and the generated force F_{pd}. It can be obtained by Eqs. (4.24)–(4.25), and Eq. (4.27) as follows:

$$
\begin{aligned}
F^r &= F^a + \underbrace{F^p + F^d}_{F_{pd}} \\
&= \underbrace{K^\alpha (x_d - x_f)}_{F^a} + \underbrace{K_p(x_d - x_f) - K_d J_f(q_f)\dot{q}_f}_{F^{pd}} \\
&= \underbrace{\left((K^\alpha_{max} - K^\alpha_{min}) \frac{(\alpha^k_i - \alpha^k_{max})}{(\alpha^k_{min} - \alpha^k_{max})} + K^\alpha_{min}\right)(x_d - x_f)}_{F^a} \\
&\quad + \underbrace{K_p \tilde{x}_e - K_d J_f(q_f)\dot{q}_f}_{F^{pd}},
\end{aligned}
\tag{4.28}
$$

where F^r indicates the force integrated variable stiffness with PD control in Cartesian space. The control law of the follower device can be expressed as follows:

$$
\begin{aligned}
u_1 &= J_f^T(q_f)F^r \\
&= J_f^T(q_f)\left(F^a + F^p + F^d\right) \\
&= J_f^T(q_f)\left(K^\alpha(x_d - x_f) + K_p(x_d - x_f) - K_d J_f(q_f)\dot{q}_f\right) \\
&= J_f^T(q_f)\left(\left((K_{max}^\alpha - K_{min}^\alpha)\frac{(\alpha_i^k - \alpha_{max}^k)}{(\alpha_{min}^k - \alpha_{max}^k)} + K_{min}^\alpha\right)(x_d - x_f)\right. \\
&\quad \left. + K_p\tilde{x}_e - K_d J_f(q_f)\dot{q}_f\right).
\end{aligned}
\tag{4.29}
$$

The control law (Eq. (4.29)) achieves the hybrid position (PD)/variable stiffness control in task space.

4.3.2 Virtual fixture

As the follower follows the leader, the position of the end–effector of the follower can be defined as

$$
P_f = (x_f, y_f, z_f)^T,
\tag{4.30}
$$

where x_f, y_f, z_f indicates the position in a xyz-coordinate system. The desired trajectory of the follower is

$$
P_{fd} = (x_{fd}, y_{fd}, z_{fd})^T.
\tag{4.31}
$$

The joint variables $(q_1^f, q_2^f, \cdots, q_n^f)$ of the follower are available through the inverse kinematics method.

Similarly, the position and desired position of the leader can be expressed as

$$
\begin{aligned}
P_l &= (x_l, y_l, z_l)^T, \\
P_{ld} &= (x_{ld}, y_{ld}, z_{ld})^T,
\end{aligned}
\tag{4.32}
$$

where P_l and P_{ld} are the actual and desired position of the leader, respectively.

Therefore, the position error of the follower end-effector is

$$P_f^e = (x_f^e, y_f^e, z_f^e)^T$$
$$= \begin{bmatrix} x_f - x_{fd} \\ y_f - y_{fd} \\ z_f - z_{fd} \end{bmatrix}. \tag{4.33}$$

Similarly, the position error of the leader end-effector is

$$P_l^e = (x_l^e, y_l^e, z_l^e)^T$$
$$= \begin{bmatrix} x_l - x_{ld} \\ y_l - y_{ld} \\ z_l - z_{ld} \end{bmatrix}, \tag{4.34}$$

where P_l^e is the position error of the leader.

The force generated by the VF is proportional to the position error of the tactile control, which is expressed as

$$F_{vf} = K_{vf} P_l^e, \tag{4.35}$$

where K_{vf} is the matrix of virtual fixture which indicates the guiding ability of VF.

4.4. Variable stiffness control

Generally, the dynamics model of a series robot manipulator can be expressed as follows:

$$M(\theta)\ddot{\theta} + C(\theta, \dot{\theta})\dot{\theta} + G(\theta) + f_{int} = \tau - f_{ext}, \tag{4.36}$$

where θ, $\dot{\theta}$, and $\ddot{\theta} \in R^{n \times 1}$ represent joint angles, velocities, and accelerations of a series robot manipulator, $M \in R^{n \times n}$ is the inertia matrix, $C(\theta, \dot{\theta})\dot{\theta} \in R^{n \times 1}$ is Coriolis and centrifugal torque, $G(\theta) \in R^{n \times 1}$ denotes gravitational force, f_{int} and f_{ext} are internal and external disturbance, respectively, and $\tau \in R^{n \times 1}$ is the motor torque vector.

Transforming this into a state-space equation, we have

$$\begin{cases} \dot{x}_1 &= x_2, \\ \dot{x}_2 &= M^{-1} \left[F(x) + d(t) + \tau \right], \\ y &= x_1, \end{cases} \tag{4.37}$$

where $x_1 = [\theta_1, \theta_2, ..., \theta_n]^T$, $x_2 = [\dot{\theta}_1, \dot{\theta}_2, ..., \dot{\theta}_n]^T$, $F(x) = -C(\theta, \dot{\theta})\dot{\theta} - G(\theta)$, and $d(t) = -f_{int} - f_{ext}$.

According to Eq. (4.37), it can be concluded that to obtain an accurate dynamic model, it is necessary to measure interference accurately. However, because of the difficulties mentioned above, we need to find another way to be practical. Thus, we design a controller combining the adaptive NN scheme with DOB technology to estimate the influence caused by dynamics uncertainties and interference.

We firstly introduce some preliminary knowledge.

$M_d \in R^{n \times n}$ is defined as a diagonal matrix with diagonal elements $m_{dii}(x) > 0$, which is part of the dynamics model and can be easily available but does not require high accuracy. Thus, there must exist an unknown matrix Δ_M for which $M_d + \Delta_M = M$. Considering Eq. (4.37), we have

$$
\begin{aligned}
M_d \dot{x}_2 &= (M - \Delta_M)\dot{x}_2 \\
&= (M - \Delta_M)M^{-1}[F(x) + d(t) + \tau] \\
&= \tau - \Delta_M M^{-1}\tau + (I - \Delta_M M^{-1})d(t) \\
&\quad + (I - \Delta_M M^{-1})F(x) \\
&= \tau + g(\tau) + r(d) + \mathbb{F}(x),
\end{aligned}
\tag{4.38}
$$

where $g(\tau) = -\Delta_M M^{-1}\tau$, $r(d) = (I - \Delta_M M^{-1})d(t)$, and $\mathbb{F}(x) = (I - \Delta_M M^{-1})F(x)$.

Lemma 2. *Generally speaking, according to the saturation degree of the system input, it is assumed that the motor torque of a normal robot or mechanical system is bounded, so $g(\tau) = -\Delta_M M^{-1}\tau$ is considered bounded.*

Lemma 3. *Suppose that the unknown internal and external disturbance $d(t)$ is bounded, that is, there is an unknown positive number constant d_1 for which $d(t) \leq d_1$.*

Next, we define the filtered tracking error f_i [36]:

$$
\begin{aligned}
f_i &= (\frac{d}{dt} + \lambda_i)^{n-1} e_i \\
&= e_i^{(n-1)} + C_{n-1}^1 \lambda_i^1 e_i^{(n-2)} + C_{n-1}^2 \lambda_i^2 e_i^{(n-3)} \\
&\quad + ... + \lambda_i^{n-1} e_i, \quad i = 1, 2, ..., n.
\end{aligned}
\tag{4.39}
$$

In Eq. (4.39), C_a^b represents mathematical combination, $\lambda = \text{diag}[\lambda_1, \lambda_2, ..., \lambda_n]$ with λ_i, $i = 1, 2, ..., n$, is a positive constant to be designed, and

$e = [e_1, e_2, ..., e_n]$ with $e_i = x_i - x_{di}$. It is easy to confirm that e_i converges to 0 as f_i converges to 0. Moreover, "n" denotes the state-space dimension. Specifically, $n = 2$ in this section as we have only two states x_1 and x_2. Therefore, we have

$$f = \dot{e} + \lambda e \qquad (4.40)$$

and

$$\begin{aligned} \dot{f} &= \ddot{e} + \lambda \dot{e} \\ &= \dot{x}_2 - \ddot{y}_d + \lambda \dot{e} \\ &= M_d^{-1}[\tau + g(\tau) + r(d) + \mathbb{F}(x)] + v, \end{aligned} \qquad (4.41)$$

where $v = -\ddot{y}_{di} + \lambda \dot{e}$.

4.4.1 Introduction of the integral Lyapunov–Krasovskii function

According to Ref. [37], we firstly introduce an integral Lyapunov–Krasovskii function as shown in Eq. (4.42), which is able to avoid the singularity problem of a controller, facilitating the design of the controller later. We have

$$V_1 = f^T M_\vartheta f, \qquad (4.42)$$

where $M_\vartheta = \int_0^1 \vartheta M_\alpha \, d\vartheta = \text{diag}\left[\int_0^1 \vartheta M_{\alpha ii}(\overline{x}_i) \, d\vartheta\right]$. The matrix $\alpha = \text{diag}[\alpha_{11}, \alpha_{22}, ..., \alpha_{nn}]$. To facilitate analysis, we define $\alpha_{11} = \alpha_{22} = ... = \alpha_{nn}$. And $\overline{x}_i = [x_1^T, x_2^T, ..., x_{n-1}^T, x_{n1}, x_{n2}, ..., \vartheta f_i + \zeta_i, ..., x_{nm}]^T \in R^{nm}$, with $\zeta = y_{di}^{(n-1)} - \xi_i$ and $\xi_i = \lambda_{i1} e_i^{n-2} + ... + \lambda_{i,n-1} e_i$. Here ϑ is an independent scalar to $\overline{x}_i$, including f and ζ. It is important to choose suitable M_d and α such that $M_{dii}\alpha_{ii} > 0$. Then we can write

$$V_1 = \sum_{i=1}^{l} f_i^2 \int_0^1 \vartheta M_{\alpha ii}\left(\overline{x}_i, \vartheta f_i + \zeta_i\right) d\vartheta. \qquad (4.43)$$

The partial derivation of Eq. (4.42) with respect to time can be obtained:

$$\dot{V}_1 = f^T \left[2M_\vartheta \dot{f} + \left(\frac{\partial M_\vartheta}{\partial f}\dot{f}\right)f + \left(\frac{\partial M_\vartheta}{\partial \overline{x}}\dot{\overline{x}}\right)f + \left(\frac{\partial M_\vartheta}{\partial \zeta}\dot{\zeta}\right)f\right], \qquad (4.44)$$

where

$$\begin{cases} \dfrac{\partial M_\vartheta}{\partial f} \dot{f} = \operatorname{diag}\left[\displaystyle\int_0^1 \vartheta \dfrac{\partial M_{\alpha ii}}{\partial f_i}\dot{f}_i d\vartheta\right], \\[3mm] \dfrac{\partial M_\vartheta}{\partial \overline{x}}\dot{\overline{x}} = \operatorname{diag}\left[\displaystyle\int_0^1 \vartheta \sum_{j=1}^n \dfrac{\partial M_{\alpha ii}}{\partial \overline{x}_j}\dot{\overline{x}}_j d\vartheta\right], \\[3mm] \dfrac{\partial M_\vartheta}{\partial \zeta}\dot{\zeta} = \operatorname{diag}\left[\displaystyle\int_0^1 \vartheta \dfrac{\partial M_{\alpha ii}}{\partial \zeta_i}\dot{\zeta}_i d\vartheta\right]. \end{cases} \tag{4.45}$$

For

$$\operatorname{diag}\left[\int_0^1 \vartheta \frac{\partial M_{\alpha ii}}{\partial f_i}\dot{f}_i d\vartheta\right] = \int_0^1 \vartheta^2 \frac{\partial M_\alpha}{\partial \vartheta}d\vartheta, \tag{4.46}$$

we can obtain

$$\begin{aligned} f^T\left(\frac{\partial M_\vartheta}{\partial f}\dot{f}\right)f &= f^T\left([\vartheta M_\alpha]\,|_0^1 - 2\int_0^1 \vartheta B_\alpha d\vartheta\right)\dot{f} \\ &= f^T M_\alpha \dot{f} - 2f^T M_\vartheta \dot{f}. \end{aligned} \tag{4.47}$$

For

$$\operatorname{diag}\left[\int_0^1 \vartheta \frac{\partial M_{\alpha ii}}{\partial \zeta_i}\dot{f}_i d\vartheta\right] = \int_0^1 \vartheta \frac{\partial M_\alpha}{\partial \vartheta}d\vartheta, \tag{4.48}$$

with $\dot{\zeta}_i = -\nu_i$, we can obtain

$$\begin{aligned} f^T\left(\frac{\partial M_\vartheta}{\partial \zeta}\dot{\zeta}\right)f &= f^T\left(-\int_0^1 \vartheta \frac{\partial M_\alpha}{\partial \vartheta}d\vartheta\right) \\ &= -f M_\alpha \nu + f^T. \end{aligned} \tag{4.49}$$

Then

$$\begin{aligned} \dot{V}_1 &= f^T M_\alpha \dot{f} - f^T M_\alpha \nu \\ &\quad + f^T\left[\left(\frac{\partial M_\vartheta}{\partial \overline{x}}\dot{\overline{x}}\right) + \int_0^1 M_\alpha \nu d\vartheta\right]. \end{aligned} \tag{4.50}$$

Replacing $\dot{f}$ with Eq. (4.41), we can obtain

$$\begin{aligned} \dot{V}_1 &= f^T M_\alpha M_d^{-1}\left[\tau + g(\tau) + r(d) + \mathbb{F}(x)\right] \\ &\quad + f^T\left[\left(\frac{\partial M_\vartheta}{\partial \overline{x}}\dot{\overline{x}}\right)f + \int_0^1 M_\alpha \nu d\vartheta\right]. \end{aligned} \tag{4.51}$$

We can deduce the following equation with the symmetric property of α, M_d, and $M_d\alpha$:

$$M_\alpha M_d^{-1} = M_d \alpha M_d^{-1} = \alpha. \tag{4.52}$$

Applying Eq. (4.52), we have

$$\dot{V}_1 = f^T \alpha \left[\tau + g(\tau) + r(d) + \mathbb{F}(x) + \Theta \right], \tag{4.53}$$

where

$$\Theta = \int_0^1 \vartheta \left(\frac{\partial M_d}{\partial \overline{x}} \dot{\overline{x}} \right) f \, d\vartheta + \int_0^1 M_d v \, d\vartheta, \tag{4.54}$$

$$\frac{\partial M_d}{\partial \overline{x}} \dot{\overline{x}} = \text{diag} \left[\sum_{j=1}^n \frac{\partial M_{dii}}{\partial \overline{x}_j} \dot{\overline{x}}_j \right], \quad i = 1, 2, ..., n. \tag{4.55}$$

4.4.2 Controller design

According to Ref. [33], the design of the controller can be divided into two parts. One is the DOB technique, which is used to estimate the internal or external unmeasurable uncertainties and disturbances; the other is the RBFNN, which is used to approximate the residual uncertainty of the robot.

Firstly, we define $\hat{D}$ as the estimation of unknown disturbance D, with

$$\hat{D} + \tilde{D} = D = g(\tau) + r(d), \tag{4.56}$$

where $\tilde{D}$ is the estimation error.

Secondly, we use RBFNN to approximate the remaining unknown dynamics uncertainties. We define

$$\mathbb{F}(x) + \tilde{D} = -W^T S(X), \tag{4.57}$$

where W is the NN weight and $S(X)$ is the RBF output, which is supposed to be bounded, $\|S(X)\| \leq S_{\max}$, with input $X = [x_1^T, x_2^T]^T$. For the RBF, we choose the Gaussian function

$$S_i(X) = \exp \left[\frac{-(X - c_i)^T (X - c_i)}{b_i^2} \right], \tag{4.58}$$

where $c_i = [c_{i1}, c_{i2}, ..., c_{i,m}]$ is the center and b_i is the width of the Gaussian function. In theory, RBFNN can smoothly approximate most continuous functions. However, in practice, there always exist errors in the process of updating weights, which we define as

$$\tilde{W} = \hat{W} - W, \qquad (4.59)$$

where $\hat{W}$ is the updating weight in real-time.

Then we can write

$$M_d \dot{x}_2 = \tau + D - WS. \qquad (4.60)$$

In order to completely express DOB, we design an auxiliary equation,

$$z = D - Kx_2, \qquad (4.61)$$

where $K = K^T > 0$ is a diagonal matrix and all its elements are positive constants.

Applying Eq. (4.60), we can calculate the derivative of z with regard to time:

$$\dot{z} = \dot{D} - K\dot{x}_2 \qquad (4.62)$$
$$= \dot{D} - KM_d^{-1} \left[\tau + D - W^T S \right]. \qquad (4.63)$$

Combining Lemma 2 and Lemma 3, we assume that the system disturbance is slowly time-varying. Thus, there must exist an constant d_l so that

$$||D|| \leq d_l. \qquad (4.64)$$

Therefore, Eq. (4.62) can be updated to be

$$\dot{\hat{z}} = -KM_d^{-1} \left[\tau + \hat{D} - \hat{W}^T S \right]. \qquad (4.65)$$

Consequently, $\hat{z}$ can be updated as in every sampling time, and the estimated value of D can be obtained as well:

$$\hat{D} = \hat{z} + Kx_2. \qquad (4.66)$$

Finally, we are able to design the control law on the basis of the DOB and RBFNN:

$$\tau = -\mathbb{G}\alpha f - \hat{D} + \hat{W}^T S(X) - \Theta. \qquad (4.67)$$

The RBFNN weight updating law is

$$\dot{\hat{W}}_i = -\Gamma_i \left[S_i(X)\alpha_{ii}f_i + \delta_i \hat{W}_i \right], \tag{4.68}$$

where $\Gamma_i \in R^n$ is one of numerous symmetric positive definite constant matrices and δ_i is one of the positive constants.

Applying the abovementioned integral Lyapunov–Krasovskii function, we take the following Lyapunov function as candidate:

$$V_2 = V_1 + \frac{1}{2}\tilde{D}^T\tilde{D} + \frac{1}{2}\sum_{i=1}^{n}\tilde{W}_i^T\Gamma_i^{-1}\tilde{W}_i. \tag{4.69}$$

Employing Eq. (4.53), Eq. (4.57), and Eq. (4.68) and substituting D, $W^T S$, and τ, the derivative of V_2 can be expressed as

$$\dot{V}_2 = f^T \left[-\mathbb{G}\alpha f + \tilde{D} + \tilde{W}S(X) \right] + \tilde{D}^T\dot{\tilde{D}} + \sum_{i=1}^{n}\tilde{W}_i^T\Gamma_i^{-1}\dot{\hat{W}}_i. \tag{4.70}$$

With the following facts,

$$\tilde{D} = D - \hat{D} = z - \hat{z} = \tilde{z}, \tag{4.71}$$

$$\begin{aligned} \dot{\tilde{D}} = \dot{\tilde{z}} = \dot{z} - \dot{\hat{z}} \\ = \dot{D} - KM_d^{-1}\left[\tilde{D} + \tilde{W}^T S \right], \end{aligned} \tag{4.72}$$

$$f^T\alpha\tilde{D} \le \frac{f\alpha\alpha f}{2} + \frac{\tilde{D}^T\tilde{D}}{2}, \tag{4.73}$$

$$\tilde{D}^T \le \frac{\tilde{D}^T\tilde{D}}{2} + \frac{||\dot{D}||^2}{2}, \tag{4.74}$$

$$\sum_{i=1}^{n}\tilde{W}_i^T S_i(X)f_i\alpha_{ii} = f^T\alpha\tilde{W}^T S(X), \tag{4.75}$$

$$\tilde{D}^T KM_d^{-1}\tilde{W}^T S(X) \le \frac{||\tilde{D}||^2}{2} + \frac{KM_d^{-1}S(X)||\tilde{W}||^2}{2}, \tag{4.76}$$

$$-\delta\tilde{W}_i^T\hat{W}_i = -\delta||\tilde{W}||_i^2 - \delta\tilde{W}_i^T, \tag{4.77}$$

$$W_i \le -\frac{\delta||\tilde{W}_i||^2}{2} + \frac{\delta||W_i||^2}{2}, \tag{4.78}$$

we derive

$$
\begin{aligned}
\dot{V}_2 \leq &-f^T \alpha \left(\mathbb{G} - \frac{1}{2} I_{n \times n} \right) \alpha f \\
&- \tilde{D}^T \left(K M_d^{-1} - 2 I_{n \times n} \right) \tilde{D} \\
&- \frac{\delta - K M_d^{-1} S_{\max}}{2} \sum_{i=1}^{n} \tilde{W}_i^T \tilde{W}_i \\
&+ \frac{d_l}{2} + \frac{\delta \|W\|^2}{2}.
\end{aligned}
\tag{4.79}
$$

To proceed, we need to choose a positive definite variable stiffness matrix $\mathbb{G}$, K, and δ in order to obtain the following inequalities:

$$
\lambda_{min} \left(\alpha \left(\mathbb{G} - \frac{1}{2} I_{n \times n} \right) \alpha \right) \geq \int_0^2 \vartheta \lambda \left(M_\alpha \right) d\vartheta,
\tag{4.80}
$$

$$
K M_d^{-1} - 2 I_{n \times n} > 0,
\tag{4.81}
$$

$$
\delta - K M_d^{-1} S_{\max} > 0.
\tag{4.82}
$$

Then, by enlarging the right-hand side of inequality (4.79), to be exact, the first three terms, we can establish

$$
\dot{V}_2 \leq -k V_2 + C,
\tag{4.83}
$$

with

$$
k = \min \left(\lambda_{\min} \left(K M_d^{-1} - 2 I_{n \times n} \right), \frac{\delta - K M_d^{-1} S_{\max}}{\sum_{i=1}^m \Gamma_i^{-1}}, 1 \right),
\tag{4.84}
$$

$$
C = \frac{d_l}{2} + \frac{\delta \|W\|^2}{2}.
\tag{4.85}
$$

In a way that is similar to solving a differential equation, inequality (4.83) can be mathematically "solved" to be

$$
V_2 \leq \left(V_2|_{t=0} - \frac{C}{k} \right) e^{-kt} + \frac{C}{k},
\tag{4.86}
$$

where t denotes time. It is apparent that the right-hand side of inequality (4.86) exponentially converges to $\frac{C}{k}$, proving V_2 is bounded; therefore, f, $\tilde{D}$, and $\hat{W}$ are bounded.

4.5. A VR-based teleoperation application

The structure of the VR-based teleoperation system proposed in this section is shown in Fig. 4.3. According to the API interface provided by Leap Motion, C++ is used to control the manipulator. The system also includes a set of real-time communication programs based on User Datagram Protocol (UDP), which is used to ensure that the operation data can be transmitted to the lower computer of the robot in real-time. The specific implementation steps are as follows:

1. Capture the human operator's gestures with Leap Motion.
2. Define the trigger mechanism.
3. Convert coordinates.
4. Generate control commands and transmit them to the robot in real-time through UDP communication.

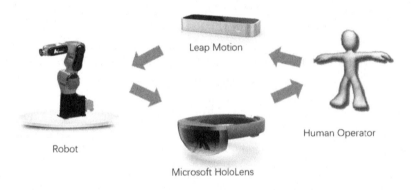

Figure 4.3 The architecture of the VR-based teleoperation system.

4.5.1 The coordinate system of Leap Motion

In order to obtain the coordinate value of the hand position captured by Leap Motion to the coordinate space of the robot, the coordinate system setting of Leap Motion should be clarified first. It is noticed that Leap Motion internal coordinates are given in millimeters. The origin of the Leap Motion coordinate system is set at the center of the top of the device, i.e., when the user moves his palm to the top center of the device, the coordinates of his palm are marked $[0; 0; 0]$. As shown in Fig. 4.4, Leap Motion uses a right-handed coordinate system.

Figure 4.4 The coordinates of Leap Motion.

Generally, transformation between coordinate systems can be achieved using the following equation:

$$x_{\text{need}} = (x_{LM} - LM_{\text{start}}) \frac{LM_{\text{range}}}{\text{need}_{\text{range}}} + \text{need}_{\text{start}}, \qquad (4.87)$$

where

$$LM_{\text{range}} = LM_{\text{end}} - LM_{\text{start}},$$
$$need_{\text{range}} = need_{\text{end}} - need_{\text{start}}. \qquad (4.88)$$

In particular, it is important to highlight an important feature called InteractionBox. This feature defines a cube area within the Leap Motion controller's view, as shown in Fig. 4.5.

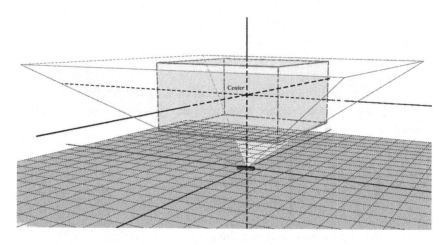

Figure 4.5 The feature named InteractionBox of Leap Motion Controller.

To get better tracking performance, the Leap Motion controller could adjust the size of the interaction box based on the field of view. However, this feature would lead to the possibility of deviations between the normalized coordinates of the point and the normalized coordinates of the point in a realistic environment.

In this work, the data collected by the Leap Motion controller will be used directly to control the industrial robot. The deviations caused by the interactive frame may make the controller for robots less effective. In some specific task environments, this can even cause damage to objects in the robot's workspace. To solve this problem, only a single fixed interactive frame object is set in this work, which is then applied to the normalization of all points within the Leap Motion's view.

4.5.2 Coordinate conversion algorithm

This section further proposes a coordinate conversion algorithm in order to transform the data collected by the Leap Motion controller into the world coordinate system of industrial robots.

At the initial moment, the coordinates of the palm in the Leap Motion controller's coordinate system are defined as $P_{\text{Leap}}\left(x_0^{\text{Leap}}, y_0^{\text{Leap}}, z_0^{\text{Leap}}\right)$, and its coordinates in the robot coordinate system are $P_r\left(x_0^r, y_0^r, z_0^r\right)$. The specific transformation is as follows:

$$\begin{cases} x_t^r = x_0^r + C_1\left(\Delta x_t^{\text{Leap}}\right), \\ y_t^r = y_0^r + C_2\left(\Delta y_t^{\text{Leap}}\right), \\ z_t^r = z_0^r + C_3\left(\Delta z_t^{\text{Leap}}\right), \end{cases} \tag{4.89}$$

where $C = (C_1, C_2, C_3)$ is an undetermined coefficient and ΔP_{Leap} is the palm displacement value.

Considering that the coordinate system of the Leap Motion controller and the industrial robot are not consistent, the above variation can be described by the following equations:

$$\begin{cases} \Delta x_t^{\text{Leap}} = z_t^{\text{Leap}} - z_0^{\text{Leap}}, \\ \Delta y_t^{\text{Leap}} = x_t^{\text{Leap}} - x_0^{\text{Leap}}, \\ \Delta z_t^{\text{Leap}} = y_t^{\text{Leap}} - y_0^{\text{Leap}}. \end{cases} \tag{4.90}$$

It should further be noted that the position data collected by the Leap Motion controller under certain environmental conditions will have some odd values that are different from the normal data. These singularities have negative impacts on the control of the robot. To solve this problem, an effective filtering algorithm needs to be designed. The specific principle of the algorithm can be expressed by the following set of equations:

$$
\begin{cases}
\Delta x_t^{\text{Leap}} = 0, \Delta x_t^{\text{Leap}} \in (-\infty, -6) \cup (6, +\infty), \\
\Delta x_t^{\text{Leap}} = 0, \Delta x_t^{\text{Leap}} \in (-0.6, 0.6),
\end{cases}
\tag{4.91}
$$

$$
\begin{cases}
\Delta y_t^{\text{Leap}} = 0, \Delta y_t^{\text{Leap}} \in (-\infty, -3) \cup (6, +\infty), \\
\Delta y_t^{\text{Leap}} = -1, \Delta y_t^{\text{Leap}} \in (-\infty, -1), \\
\Delta y_t^{\text{Leap}} = 1, \Delta y_t^{\text{Leap}} \in (1, \infty),
\end{cases}
\tag{4.92}
$$

$$
\begin{cases}
\Delta z_t^{\text{Leap}} = 0, \Delta z_t^{\text{Leap}} \in (-\infty, -5.6) \cup (5.6, +\infty), \\
\Delta z_t^{\text{Leap}} = 0, \Delta x_t^{\text{Leap}} \in (-1, 1).
\end{cases}
\tag{4.93}
$$

4.6. Experimental case study

4.6.1 Experimental results of variable stiffness control and tremor suppression

In order to verify the performance of the proposed methods in Section 4.2, two experiments are introduced in this part.

The root mean square error (RMSE) is used to quantify the performance. It is defined as

$$
RMSE = \sqrt{\frac{\sum_{i=1}^{N} (x(i) - \hat{x}(i))^2}{N}},
\tag{4.94}
$$

where $x(i)$ and $\hat{x}(i)$ are the follower device trajectory sample and the follower device trajectory sample, respectively, and N is the number of samples.

(1) Experimental setup

In this experiment, a PD controller is used to control the telerobot. The sampling period T is set as 5 s, the sampling time $t = 0.01$ s, and the nonlinear shape factor $A = -0.6891$. The controller parameters are $K_l = 1000$, $D_l = 50$, $K_f = 30$, and $D_f = 40$. By using the simplex grid search method, the parameters for the SVM are $C = 98.3355$, $\sigma = 1.3707e^{-4}$, $q = 0$, $\alpha = 1$, $d = 0.5$, $N = 500$.

(2) Tremor attenuation

Table 4.1 shows the tracking performance in the case of tremor or tremor attenuation. The value of the RMSE with tremor attenuation is smaller than that with tremor, which indicates that the proposed method is effective in attenuating tremor. It can be concluded that the proposed SVM method achieves better performance for teleoperation.

Table 4.1 RMSE of teleoperation in the case of tremor and tremor attenuation.

Mode	Direction	RMSE
With tremor	x-direction	0.0416
With tremor attenuation	x-direction	0.0228

(3) Teleoperational control strategy

In order to verify the effectiveness of the proposed variable stiffness method, we conducted flutter attenuation teleoperation tests in three modes: high stiffness, low stiffness, and variable stiffness.

Table 4.2 shows the comparisons among high stiffness, low stiffness, and variable stiffness. The performance of teleoperation in variable gain mode is better than that in high stiffness and low stiffness modes. Compared with the other two modes, the proposed method based on variable stiffness has the best performance in the case of trajectory tracking. It is seen that the proposed method performs best in terms of accuracy, effectiveness, and stability.

Table 4.2 Comparison among high stiffness, low stiffness, and variable stiffness.

Mode	Direction	RMSE
High stiffness	x-/y-/z-direction	0.0408/0.1768/0.1390
Low stiffness	x-/y-/z-direction	0.0285/0.1062/0.0639
Variable stiffness	x-/y-/z-direction	0.0118/0.0629/0.0396

4.6.2 Experimental results of variable stiffness control and virtual fixture

(1) Tracking experiment

Tracking experimental results are shown in Fig. 4.6. In Fig. 4.6(a), due to different initial positions in task space, the follower does not track the leader in the first 3 seconds. In the last 7 seconds, the follower almost completely follows the trajectories of the leader. The tracking error curve of the follower is shown in Fig. 4.6(c). In Figs. 4.6(b)–4.6(d), the follower can closely follow the leader after 0–1 s in the x-/y-direction. It is clear that the integrated control method shows better performance in the trajectory tracking experiment than the PD control method.

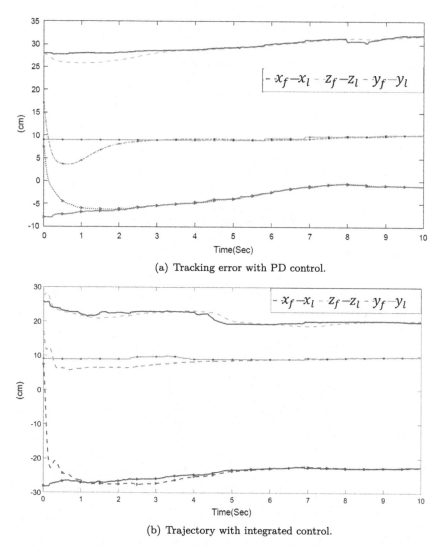

(a) Tracking error with PD control.

(b) Trajectory with integrated control.

Figure 4.6 Contrast between PD control and integrated control.

As shown in Table 4.3, the RMSE value of the integrated control mode is smaller than that of the PD control mode. In other words, the tracking

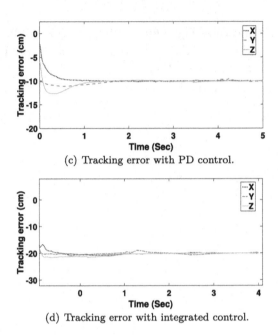

(c) Tracking error with PD control.

(d) Tracking error with integrated control.

Figure 4.6 (*continued*)

performance of the proposed integrated control method is better than that of the PD control method.

Table 4.3 Comparison of RMSE values among PD control and integrated control.

Direction	PD control	Integrated control
x-direction	0.0176	0.0108
y-direction	0.0093	0.0081
z-direction	0.0173	0.0160

(2) Virtual fixture experiment

In this experiment, two typical trajectories of the follower device end-effector in task space are presented to verify the performance of the proposed method by using the integrated control method.

The triangle trajectory performance of the follower in the operating space is shown in Fig. 4.7. Fig. 4.7(a) reveals that the follower cannot follow the leader exactly. Figs. 4.7(b)–4.7(e) show the whole interaction process between the human operator and the telerobot. Comparing Figs. 4.7(c) and 4.7(a), it is clear that the tracking error is relatively small when using

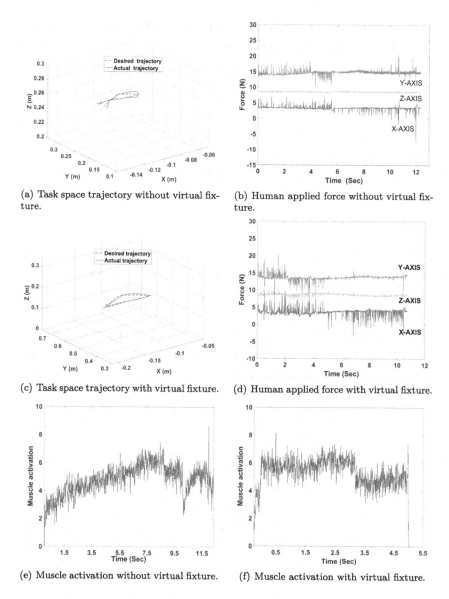

(a) Task space trajectory without virtual fixture.

(b) Human applied force without virtual fixture.

(c) Task space trajectory with virtual fixture.

(d) Human applied force with virtual fixture.

(e) Muscle activation without virtual fixture.

(f) Muscle activation with virtual fixture.

Figure 4.7 Task space performance in angle trajectory with and without VF.

VF, indicating that the operation performance of the follower with VF is superior to that without VF. Experimental observations show that the time spent on the triangle trajectory task without VF is 12.33 s, while the completion time is 10.72 s in the case of VF.

4.6.3 Experimental results of disturbance observer-enhanced variable stiffness controller

This section demonstrates the feasibility of the controller proposed in Section 4.4. The tracking performance is verified through experiments and simulations.

In this verification, we utilized the first two joints of the Baxter robot ("left_s0" and "left_s1"), while other joints are not considered as part of the control object but viewed as a payload attached to the joint "left_s1" as well as disturbance.

(1) Performance of different controllers

In this part, we collected trajectories using the proposed scheme. These trajectories were collected when human operators extended their arms and drew irregular circles with their hands, as shown in Fig. 4.8(b). The generated trajectories are shown in Fig. 4.9(a). Then, we used them to track the above two joints of the Baxter robot by utilizing the PD controller to force the other five joints to maintain the zero position. It should be pointed out that the coupling dynamics of the last five links can be regarded as a time-varying random disturbance to the first two joints.

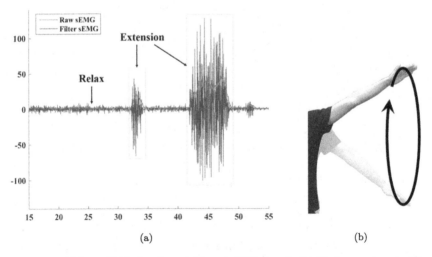

Figure 4.8 (a) Raw sEMG signals and filtered sEMG signals. (b) Motion captured with operator stretching out his arm and drawing circles with his hand.

To compare our proposed controller with others, we referred to a traditional PD controller and a PD controller integrated with an NDOB proposed by Wenhua Chen in Ref. [27].

The selected PD controller is $\tau = -K_{1d}\dot{e} - K_{1p}e$, with $K_{1d} = \text{diag}[30, 29]$ and $K_{1p} = \text{diag}[225, 220]$.

The chosen PD controller with NDOB is

$$
\begin{aligned}
\tau &= -K_{2d}\dot{e} - K_{2p}e - \hat{d}, \\
\hat{d} &= z + p(\theta, \dot{\theta}), \\
\dot{z} &= -L(\theta, \dot{\theta})z + L(\theta, \dot{\theta})(G(\theta, \dot{\theta}) - T - p(\theta, \dot{\theta})), \\
G(\theta, \dot{\theta}) &= C_{\text{NDOB}} + G_{\text{NDOB}}, \\
L(\theta, \dot{\theta}) &= c \begin{bmatrix} 1 & 0 \\ 1 & 1 \end{bmatrix} M_{\text{NDOB}}^{-1}, \\
p(\theta, \dot{\theta}) &= c \begin{bmatrix} \dot{\theta}_1 \\ \dot{\theta}_1 + \dot{\theta}_2 \end{bmatrix},
\end{aligned}
\tag{4.95}
$$

where $c = 0.01$, $K_{2d} = \text{diag}[22, 20]$, $K_{2p} = \text{diag}[215, 215]$, and M_{NDOB}, C_{NDOB}, and G_{NDOB} are the first three links dynamics model of the Baxter robot, which can be nominally obtained [38].

For the proposed controller, M_d was chosen as $\text{diag}[0.05, 0.05]$, implying that we knew little about the controlled object model. Other variables are shown in Table 4.4.

For the three controllers mentioned above, in order to show the function of RBFNN in the proposed controller, we simply remove the RBF term (the third item in Eq. (4.67)) in the proposed controller.

Table 4.4 Variables of the proposed controller.

Parameter	Value	Parameter	Value
Sampling rate	100 Hz	Node centers	Combination of θ_1 (0.1, 0.9), θ_2 (−0.9, −0.1), $\dot{\theta}_1$ (−0.24, 0.24), and $\dot{\theta}_2$ (−0.24, 0.24)
Control rate	100 Hz	λ	$\text{diag}[17.5, 16]$
No. of RBF nodes	16	$\mathbb{G}$	$\text{diag}[17.5, 16]$
Node variance	1	K	$\text{diag}[0.1, 0.1]$
Initial weight	0		

The experimental results are shown in Fig. 4.9. Figs. 4.9(a) and 4.9(b) show the desired and real-time recorded trajectories and the tracking errors

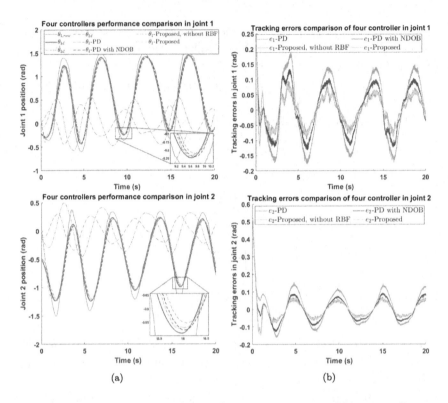

Figure 4.9 Experimental results of the performance comparison of three controllers.

of the four controllers. Figs. 4.9(c) and 4.9(d) depict the control outputs and the norm of RBFNN weights.

Following these results, we can conclude that (i) for the proposed controller, the tracking trajectory and error fluctuations will be more severe if the RBFNN item is cut off and (ii) employing the proposed controller, tracking errors remain minimal most of the time. Hence, we can conclude that (i) if there is no RBFNN, the system will become unstable, that is to say, the RBFNN integrated with the proposed controller can reduce the influence of model uncertainty, and (ii) the proposed controller can perform well even in the serious situation of difficult to measure disturbance, better than the PD controller and PD controller with NDOB. The feasibility of the proposed controller scheme combining the RBFNN algorithm with DOB technology is verified by the simulation to reduce the adverse effects of model uncertainty and disturbance.

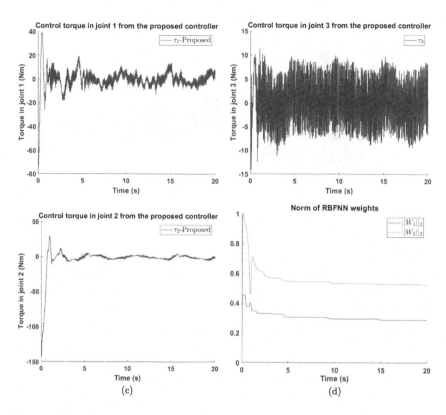

Figure 4.9 (*continued*)

(2) Effectiveness of variable stiffness control

Imagine we need a robot to save a pet. The robot needs to hold it and put it in a safe box, but the animal is afraid and tries to escape from the end-effector of the robot. In this case, we need a strategy to make the end-effector be close to the expected position of unknown and fast time-varying interference. Therefore, we propose a variable stiffness control scheme. The following study demonstrates its effectiveness.

To verify the anti-disturbance performance of the proposed variable stiffness control scheme, we firstly collect the time series of sEMG signals, which was employed offline, simply supposing that the operator began to tense or relax his arm muscles when he realized the teleoperation situation required him to modify the control gain. Secondly, we generate disturbances of different frequencies and amplitudes by the method shown in Fig. 4.10, in which the sixth joint driving the Baxter robot follows a sine

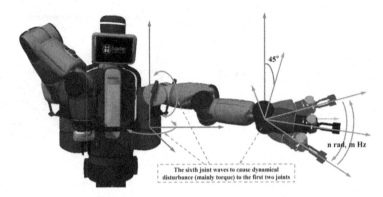

Figure 4.10 Disturbance is created through driving the sixth joint to track a sine curve, while the first two joints are controlled to stay in zero position.

wave: $\theta_{6d} = n\sin(2m\pi t)$. Thirdly, we try to maintain the zero position of the first two joints of the Baxter robot (in Gazebo) under different interference conditions by using the developed variable stiffness method.

In this part, the parameters we utilized are the same as those listed in Table 4.4 except for those in Table 4.5.

Table 4.5 Parameters utilized in variable stiffness control tests.

Parameter	Value
Node centers	Combination of $\theta_1(-0.4, 0.4)$, $\theta_2(-0.4, 0.4)$, $\dot{\theta}_1(-0.56, 0.56)$, and $\dot{\theta}_2(-0.56, 0.56)$
$\mathbb{G}$	$\mathbb{G}_{\max} = \text{diag}[30, 28]$, $\mathbb{G}_{\min} = \text{diag}[15.5, 15]$
A	-0.04
N	1
(n, m) in Fig. 4.10	$(\pm 0.8, 1)$

To create disturbance, we forced the sixth joint to track a sine wave with different frequencies in a fixed direction in Fig. 4.10. Such direction ensures that this kind of dynamical disturbance is generated by the sixth joint and influences the first two joints.

Then we conducted the simulations under different circumstances. One result is shown in Fig. 4.11, while the others are shown in Table 4.6.

In every panel of Fig. 4.11, the results are divided into five stages, which are shown by two alternating gray background colors.

Stage 1 is from 0 s to 5 s and the first two joints are free from fast time-varying disturbance. They normally stay in the zero position (Fig. 4.11(a)). We can see the controller functions well in such peaceful environment,

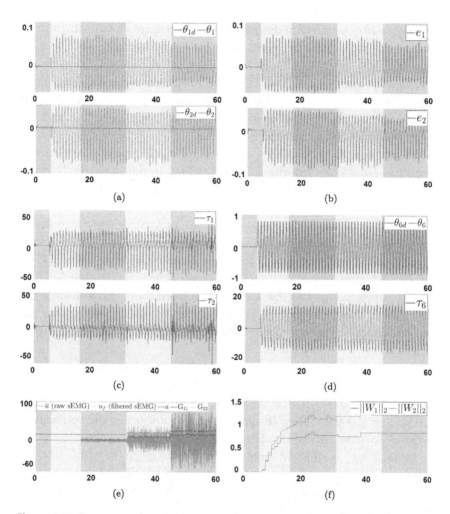

Figure 4.11 Experimental result of the effectiveness of variable stiffness control based on sEMG.

though the "attached" static disturbance (payload of the last five joints) never disappears.

Stage 2 is from 5 s to 15 s. When the fast time-varying disturbance starts, the sEMG is not yet employed. We can see the influence of the movement of joint 6 on joint 1 and joint 2: the tracking errors of joint 1 and joint 2 increase up to around ±0.6 rad and ±0.5 rad, respectively (Fig. 4.11(a) and Fig. 4.11(b)). Though the RBFNN starts working (Fig. 4.11(f)), the DOB is of no use to this kind for fast time-varying and asymmetrical disturbance.

The asymmetry property of disturbance can be reflected by the unsymmetrical errors, which mainly results from the particular movement limit of joint 4.

Stage 3 is from 15 s to 30 s. The sEMG begins to work and the control gains G_{11} and G_{22} are lightly modified because (suppose) the operator still does not notice the disturbance (Fig. 4.11(e)) and no change in the tracking errors of joint 1 and joint 2 can be noticed.

Stage 4 is from 30 s to 45 s. Now (suppose) the operator realizes the disturbance and keeps his muscle tensed at first. In this stage, the tracking errors of joint 1 and joint 2 apparently decrease (Fig. 4.11(b)) as the control stiffness increases (Fig. 4.11(e)).

Stage 5 is from 45 s to 60 s, where the movement of joint 6 keeps going but the control stiffness moves to another higher state, which results in the smallest errors in joint 1 and joint 2 among all stages (Fig. 4.11(a) and Fig. 4.11(b)).

The panels in Fig. 4.11 that are not mentioned above also provide some information about the process, especially Fig. 4.11(c), where we can see that the controller outputs (control torques) increase as the sEMG signal increases.

The RMSE values are also computed: $RMSE_{1,1}$ (joint 1, 17.5–27.5 s) $= 0.00127$, $RMSE_{1,2}$ (joint 1, 32.5–42.5 s) $= 0.00104$, $RMSE_{1,3}$ (joint 1, 47.5–57.5 s) $= 0.00079$, $RMSE_{2,1} = 0.00120$, $RMSE_{2,2} = 0.00103$, and $RMSE_{2,3} = 0.00080$. They, in another way, contribute to the verification of the effectiveness of the variable gain control scheme.

In Table 4.6, another 12 simulation results are given, in which all conditions are the same as in the aforementioned simulation except for those marked out in the second column. Similarly, all RMSE results reveal that as the operator tenses his forearm muscles, the tracking errors of joint 1 and joint 2 tend to decrease, irrespective of the disturbances, the control gain intervals, or the amplitudes of the operator's arm muscle sEMG (note that these sEMG series were obtained when the operator regularly relaxed/tensed his arm muscles; see Fig. 4.11(e)).

Therefore, the proposed variable stiffness control scheme is practical and able to resist fast time-varying disturbances.

4.7. Conclusion

In this chapter, several new types of EMG signal-based variable stiffness control based on tremor suppression, and VF human–robot interactive

Table 4.6 Variable stiffness control tests results.

No.	Changed condition	RMSE results		
		$RMSE_{1,1}$	$RMSE_{1,2}$	$RMSE_{1,3}$
		$RMSE_{2,1}$	$RMSE_{2,2}$	$RMSE_{2,3}$
1	$m = 1.1$	0.00135	0.00118	0.00097
		0.00126	0.00111	0.00092
2	$m = 0.9$	0.00111	0.00082	0.00064
		0.00105	0.00081	0.00067
3	$m = 0.8$	0.00082	0.00059	0.00047
		0.00074	0.00056	0.00046
4	$n = \pm 1$	0.00125	0.00099	0.00084
		0.00119	0.00096	0.00086
5	$n = \pm 0.9$	0.00126	0.00103	0.00086
		0.00122	0.00102	0.00085
6	$n = \pm 0.7$	0.00125	0.00100	0.00081
		0.00119	0.00098	0.00082
7	$\mathbb{G}_{max} = \mathbb{G}_{max} + 2$	0.00131	0.00099	0.00080
		0.00125	0.00096	0.00082
8	$\mathbb{G}_{min} = \mathbb{G}_{min} - 2$ $\mathbb{G}_{max} = \mathbb{G}_{max} + 2$	0.00140	0.00109	0.00082
		0.00131	0.00106	0.00087
9	$\mathbb{G}_{min} = \mathbb{G}_{min} - 2$ $\mathbb{G}_{max} = \mathbb{G}_{max} - 2$	0.00137	0.00112	0.00087
		0.00130	0.00110	0.00088
10	sEMG series 1	0.00131	0.00124	0.00096
		0.00125	0.00120	0.00096
11	sEMG series 2	0.00130	0.00120	0.00097
		0.00126	0.00116	0.00095
12	sEMG series 3	0.00129	0.00105	0.00084
		0.00123	0.00103	0.00085

control are proposed for a teleoperation robot system. Focusing on the characteristics of the tremor signal, this chapter a linear learning weight function is designed. In the process of tremor filtering, the SVM filter can effectively use small samples to filter the tremor, and the variable stiffness control method can achieve personalized control according to muscle activity. The combination of tremor filtering and personalized control methods improves the control performance of remote operation. In addition, combined with the hybrid control of variable stiffness control and the VF program, the activity of the operator's hand muscles can be effectively adjusted, providing a natural and safe HRI for the operator and improving the operating level. Additionally, in this chapter a controller with a variable stiffness scheme is designed. By utilizing the proposed controller, integrating DOB technology, the RBFNN algorithm, and the variable stiffness strategy, the trajectory tracking performance of the robot manipulator is better than that of previously mentioned controllers. Model dynamics uncertainty problems are out of account, since the proposed combination of DOB and RBFNN reduces the reliability of dynamic nonlinear models. Finally, the experimental results show that in the case of tremor filtering, the teleoperation robot system can improve the tracking control performance, and the control stiffness of the teleoperation system can adapt to the change of the operator's muscle activity. The methods of tremor filtering–individual control and hybrid control–VF in this chapter provide a good foundation for user-friendly and natural HRI of teleoperation robot systems. Experimental and simulation results demonstrate that the variable stiffness control scheme is practical and effective in resisting fast time-varying disturbances to some extent, enabling the telerobot manipulator to copy human arm motions and to resist low- and high-frequency disturbances.

References

[1] Joseph Yan, Septimiu E. Salcudean, Teleoperation controller design using H/sub /spl infin//-optimization with application to motion-scaling, IEEE Transactions on Control Systems Technology 4 (3) (1996) 244–258.
[2] Xian Yang, Chang-Chun Hua, Jing Yan, Xin-Ping Guan, A new master–slave torque design for teleoperation system by ts fuzzy approach, IEEE Transactions on Control Systems Technology 23 (4) (2014) 1611–1619.
[3] Jacob Rosen, Blake Hannaford, Mark P. MacFarlane, Mika N. Sinanan, Force controlled and teleoperated endoscopic grasper for minimally invasive surgery-experimental performance evaluation, IEEE Transactions on Biomedical Engineering 46 (10) (1999) 1212–1221.
[4] Yiming Jiang, Chenguang Yang, Shi-lu Dai, Beibei Ren, Deterministic learning enhanced neutral network control of unmanned helicopter, International Journal of Advanced Robotic Systems 13 (6) (2016) 1729881416671118.
[5] Sivanagaraja Tatinati, Kalyana C. Veluvolu, Wei Tech Ang, Multistep prediction of physiological tremor based on machine learning for robotics assisted microsurgery, IEEE Transactions on Cybernetics 45 (2) (2014) 328–339.
[6] Sivanagaraja Tatinati, Kalyana C. Veluvolu, Sun-Mog Hong, Win Tun Latt, Wei Tech Ang, Physiological tremor estimation with autoregressive (AR) model and Kalman filter for robotics applications, IEEE Sensors Journal 13 (12) (2013) 4977–4985.
[7] Kalyana C. Veluvolu, Sivanagaraja Tatinati, Sun-Mog Hong, Wei Tech Ang, Multistep prediction of physiological tremor for surgical robotics applications, IEEE Transactions on Biomedical Engineering 60 (11) (2013) 3074–3082.

[8] Louis B. Rosenberg, Virtual fixtures: Perceptual tools for telerobotic manipulation, in: Proceedings of IEEE Virtual Reality Annual International Symposium, 1993, pp. 76–82.

[9] Mark A. Fehlberg, Ilana Nisky, Andrew J. Doxon, William R. Provancher, Improved active handrest performance through use of virtual fixtures, IEEE Transactions on Human-Machine Systems 44 (4) (2014) 484–498.

[10] Brian C. Becker, Robert A. MacLachlan, Louis A. Lobes, Gregory D. Hager, Cameron N. Riviere, Vision-based control of a handheld surgical micromanipulator with virtual fixtures, IEEE Transactions on Robotics 29 (3) (2013) 674–683.

[11] Minsik Hong, Jerzy W. Rozenblit, A haptic guidance system for computer-assisted surgical training using virtual fixtures, in: 2016 IEEE International Conference on Systems, Man, and Cybernetics (SMC), IEEE, 2016, pp. 002230–002235.

[12] Yaser Maddahi, Kourosh Zareinia, Nariman Sepehri, An augmented virtual fixture to improve task performance in robot-assisted live-line maintenance, Computers and Electrical Engineering 43 (2015) 292–305.

[13] Mario Selvaggio, Gennaro Notomista, Fei Chen, Boyang Gao, Francesco Trapani, Darwin Caldwell, Enhancing bilateral teleoperation using camera-based online virtual fixtures generation, in: 2016 IEEE/RSJ International Conference on Intelligent Robots and Systems (IROS), 2016, pp. 1483–1488.

[14] Camilo Perez Quintero, Masood Dehghan, Oscar Ramirez, Marcelo H. Ang, Martin Jagersand, Flexible virtual fixture interface for path specification in tele-manipulation, in: 2017 IEEE International Conference on Robotics and Automation (ICRA), 2017, pp. 5363–5368.

[15] Jian Zhong Qiao, Hao Wu, Yukai Zhu, Jianwei Xu, Wenshuo Li, Anti-disturbance iterative learning tracking control for space manipulators with repetitive reference trajectory, Assembly Automation (2019).

[16] Yanchao Sun, Liangliang Chen, Hongde Qin, Distributed adaptive fuzzy tracking algorithms for multiple uncertain mechanical systems, Assembly Automation (2019).

[17] Liang Li, Ziyu Chen, Yaobing Wang, Xiaodong Zhang, Ningfei Wang, Robust task-space tracking for free-floating space manipulators by cerebellar model articulation controller, Assembly Automation (2019).

[18] Zidong Wang, Hong Qiao, Keith J. Burnham, On stabilization of bilinear uncertain time-delay stochastic systems with Markovian jumping parameters, IEEE Transactions on Automatic Control 47 (4) (2002) 640–646.

[19] Zidong Wang, Hong Qiao, Robust filtering for bilinear uncertain stochastic discrete-time systems, IEEE Transactions on Signal Processing 50 (3) (2002) 560–567.

[20] Kouhei Ohnishi, Masaaki Shibata, Toshiyuki Murakami, Motion control for advanced mechatronics, IEEE/ASME Transactions on Mechatronics 1 (1) (1996) 56–67.

[21] Emre Sariyildiz, Kouhei Ohnishi, Stability and robustness of disturbance-observer-based motion control systems, IEEE Transactions on Industrial Electronics 62 (1) (2015) 414–422.

[22] Zi Jiang Yang, Y. Fukushima, Pan Qin, Decentralized adaptive robust control of robot manipulators using disturbance observers, IEEE Transactions on Control Systems Technology 20 (5) (2012) 1357–1365.

[23] Hoi Wai Chow, N.C. Cheung, Disturbance and response time improvement of submicrometer precision linear motion system by using modified disturbance compensator and internal model reference control, IEEE Transactions on Industrial Electronics 60 (1) (2013) 139–150.

[24] Wataru Iida, Kouhei Ohnishi, Reproducibility and operationality in bilateral teleoperation, in: The IEEE International Workshop on Advanced Motion Control, 2004, pp. 217–222.

[25] Kwang Sik Eom, Il Hong Suh, Wan Kyun Chung, Disturbance observer based path tracking control of robot manipulator considering torque saturation, Mechatronics 11 (3) (2001) 325–343.

[26] Mou Chen, Wen-Hua Chen, Sliding mode control for a class of uncertain nonlinear system based on disturbance observer, International Journal of Adaptive Control and Signal Processing 24 (1) (2010) 51–64.

[27] Wen-Hua Chen, Donald J. Ballance, Peter J. Gawthrop, John O'Reilly, A nonlinear disturbance observer for robotic manipulators, IEEE Transactions on Industrial Electronics 47 (4) (2000) 932–938.

[28] Amin Nikoobin, Reza Haghighi, Lyapunov-based nonlinear disturbance observer for serial n-link robot manipulators, Journal of Intelligent and Robotic Systems 55 (2–3) (2009) 135–153.

[29] Mou Chen, WenHua Chen, QinXian Wu, Adaptive fuzzy tracking control for a class of uncertain MIMO nonlinear systems using disturbance observer, Science in China, Series F, Information Sciences 57 (1) (2014) 1–13.

[30] Zi Jiang Yang, Hiroshi Tsubakihara, S. Kanae, K. Wada, A novel robust nonlinear motion controller with disturbance observer, IEEE Transactions on Control Systems Technology 16 (1) (2006) 137–147.

[31] Arash Ajoudani, Marco Gabiccini, N. Tsagarakis, Alin Albu-Schäffer, Antonio Bicchi Teleimpedance, Exploring the role of common-mode and configuration-dependent stiffness, in: 2012 12th IEEE-RAS International Conference on Humanoid Robots (Humanoids 2012), IEEE, 2012, pp. 363–369.

[32] Zhijun Li, Yu Kang, Zhiye Xiao, Weiguo Song, Human–robot coordination control of robotic exoskeletons by skill transfers, IEEE Transactions on Industrial Electronics 64 (6) (2016) 5171–5181.

[33] Longbin Zhang, Zhijun Li, Chenguang Yang, Adaptive neural network based variable stiffness control of uncertain robotic systems using disturbance observer, IEEE Transactions on Industrial Electronics 64 (3) (2016) 2236–2245.

[34] Chenguang Yang, Xingjian Wang, Zhijun Li, Yanan Li, Chun-Yi Su, Teleoperation control based on combination of wave variable and neural networks, IEEE Transactions on Systems, Man, and Cybernetics: Systems 47 (8) (2016) 2125–2136.

[35] Zhi Liu, Jing Luo, Liyang Wang, Yun Zhang, CL Philip Chen, Xin Chen, A time-sequence-based fuzzy support vector machine adaptive filter for tremor cancelling for microsurgery, International Journal of Systems Science 46 (6) (2015) 1131–1146.

[36] Jean-Jacques E. Slotine, Weiping Li, et al., Applied Nonlinear Control, vol. 199, Prentice Hall, Englewood Cliffs, NJ, 1991.

[37] Zhijun Li, Ziting Chen, Jun Fu, Changyin Sun, Direct adaptive controller for uncertain MIMO dynamic systems with time-varying delay and dead-zone inputs, Automatica 63 (2016) 287–291.

[38] Alex Smith, Chenguang Yang, Chunxu Li, Hongbin Ma, Lijun Zhao, Development of a dynamics model for the Baxter robot, in: 2016 IEEE International Conference on Mechatronics and Automation, IEEE, 2016, pp. 1244–1249.

CHAPTER FIVE

Shared control for teleoperation

This chapter is mainly focused on shared control for teleoperation. First, a brief review of the concept and the applications of shared control will be given in Section 5.1. Second, several commonly used applications of shared control, like collision avoidance control, electroencephalography (EEG)-based shared control, mixed reality (MR)-based user interactive path planning, surface electromyography (sEMG)-based shared control, and artificial potential field (APF)-based hybrid shared control, will be introduced. Experimental results will be analyzed in Section 5.7. Finally, we will give a summary of this chapter in Section 5.8.

5.1. Introduction

Shared control, as a combination of purely autonomous and tele-operated control modes [1], can effectively assist in task execution and decrease workload demands on the human operator [2]. In application environments with large delays, the concept of shared control can also be considered as humans and automatic devices working together to achieve a common goal. In addition, in a shared control system, the operator specifies the task and then the automatic machine completes it by itself.

Shared control has proved to be an efficient method for designing intuitive robotic teleoperation interfaces for human operators, which could reduce human operators' workload when they carry out complex tasks [3]. Shared control in teleoperation systems makes it possible to share the available degrees of freedom of a robotic system between the operator and an autonomous controller, to facilitate the task for the human operator and improve the overall efficiency of the system. Taking a robotic cutting example, it has high dexterity and safety requirements. For example, Prada and Payandeh used geometric virtual fixtures to assist with robot cutting. Besides, the shared control strategy has been employed in obstacle avoidance, in which the human operator only needs to consider the motion of the end-effector of the manipulator [4]. Moreover, combining the shared control method with the EMG sensor has been proposed to enable human to teleoperate a mobile robot and achieve obstacle avoidance simultaneously [5]. The force feedback based on muscle activation can be transmitted to the human to update their control intention with predictability. In Ref. [6], a passive task-prioritized shared control method for remote telemanipulation of redundant robots was proposed. Haptic feedback and guidance have been shown to play a significant and promising role in shared control applications. Haptic cues can be used to increase situation awareness and to effectively steer the human operator towards the safe execution of some tasks.

Shared control schemes are often combined with other control methods in practice. For example, in Ref. [7], shared control with an adaptive servo method is presented to assist disabled people to complete a transport task which integrates a tracking controller and an obstacle avoidance controller. In a complex environment, outputs of a compliance motion control and autonomous navigation control are combined to form the inputs of a shared controller [8]. Furthermore, force feedback of mobile robots is often used to help the human partner to improve the perception of environments for enhancing the operation skills [9,10]. Obstacle avoidance is one of the most important tasks in the research area of mobile robots. When a mobile robot follows the commands of a human partner to a target position, it must avoid the obstacles autonomously at the same time.

5.2. Collision avoidance control

As is well known, the manipulator is an essential part of robotics that has received increasing attention in the research and industry communities. It has outstanding performance in dangerous tasks, human–robot interaction (HRI), etc. In order to ensure the safety of humans and manipulators, collision avoidance is considered as a crucial function which enables the manipulator to avoid collisions with known obstacles.

A feasible method is embedding an automatic collision avoidance mechanism into the shared control framework and the teleoperation system. In this manner, on the one hand, the shared control can decrease the workload of the human operator, and on the other hand, the automatic collision avoidance method allows the operator to focus on the manipulation of the end–effector rather than caring about whether the manipulator will collide with its surroundings.

The aforementioned control system is achieved at the kinematic and dynamic levels to guarantee the performance of the manipulator (Fig. 5.1). At the kinematic level, automatic collision avoidance is achieved by exploiting the joint space redundancy, and the goal of the design at the kinematic level is to generate a reference trajectory in the joint space for the end-effector of the manipulator to follow the operator's commands and simultaneously achieve collision avoidance. At the dynamic level, a radial basis function neural network (RBFNN) is developed to compensate for the effect caused by internal and external uncertainties such as unknown payload. The goal of the design at the dynamic level is to ensure the reference trajectory can be tracked satisfying a specified performance requirement in the presence of uncertainties.

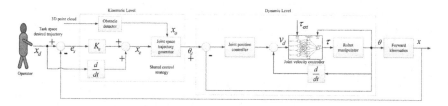

Figure 5.1 System diagram.

One of the feasible experiments is illustrated in Fig. 5.2. An operator teleoperates the Baxter robot's manipulator using an Omni joystick connected to the leader computer and the Baxter robot's manipulator is connected to the follower computer. A Kinect sensor is used to detect obstacles in the robot's surrounding environment.

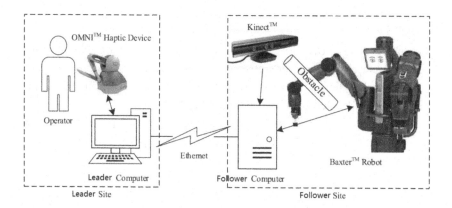

Figure 5.2 Components used in the experimental setup.

5.2.1 Kinematic level

(1) Identification of collision points

In this experiment, a self-identification method has been developed based on the 3D point cloud and the robot skeleton, The point cloud is over-segmented into superpixels based on the continuous K-means clustering method. The superpixels on the robot are then identified using the robot skeleton obtained from the forward kinematics. According to the segmented point cloud, a simplified 3D model of the robot is generated automatically in real-time which is drawn by red spheres [11].

As shown in Fig. 5.3, p_{cr} and p_{co} denote the collision points in the obstacle and the 3D model of the robot. The black lines from p_{cr} to p_{co} indicate that the measured distance between the robot's manipulator and the obstacle.

Figure 5.3 Obstacle detection.

(2) Collision avoidance

The control goal at the kinematics level is to make the manipulator's end-effector precisely follow the reference trajectory in the Cartesian space commanded by the operator and simultaneously avoid collision automatically. It can be divided into two small goals.

The first goal is to find the joint velocities of the manipulator to follow the reference trajectory. The kinematics of the manipulator are given by

$$\dot{x}_e = J_e \dot{\theta}, \tag{5.1}$$

where $\dot{\theta}$ is the joint velocities, $\dot{x}_e$ is the end-effector velocity, and J_e is the end-effector Jacobian matrix.

In order to achieve the first goal, a closed-loop kinematics control law is designed as follows:

$$\dot{x}_e = \dot{x}_d + K_e e_x, \tag{5.2}$$

where the position error e_x can be calculated by $e_x = x_d - x_e$, x_d denotes the desired position, x_e is the actual position, and K_e is a positive definite gain matrix which is chosen by the designer.

The second goal is to avoid the potential collision with minimal effect on the first goal. Because the manipulator of the Baxter robot is redundant, we can achieve this goal by exploiting the kinematic redundancy mechanism of the manipulator in the joint space. Therefore, a general inverse kinematics solution can be described as

$$\dot{\theta} = J^{\dagger}\dot{x} + (I - J^{\dagger}J)z, \tag{5.3}$$

where $J^{\dagger}$ is the pseudo–inverse of the Jacobian matrix, which is defined as $J^{\dagger} = J^T(JJ^T)^{-1}$, and z is a vector which is used for collision design [12].

The desired avoiding velocity $\dot{x}_o$ can make the collision point $\boldsymbol{p}_{cr}$ move toward the direction opposite to the obstacle velocity's direction (Fig. 5.4). It must satisfy the following kinematic constraint:

$$\dot{x}_o = J_o\dot{\theta}, \tag{5.4}$$

where J_0 is the Jacobian matrix of the collision point $\boldsymbol{p}_{cr}$. In order to reduce the difficulty in computation, J_o can be simplified as follows:

$$J_o = [J_{e1}, ..., J_{el}, 0, ..., 0], \tag{5.5}$$

where l is the number of joints which are above the potential collision point $\boldsymbol{p}_{cr}$.

It is obvious that the collision point $\boldsymbol{p}_{cr}$ should move faster when the obstacle is closer to $\boldsymbol{p}_{cr}$, so $\dot{x}_o$ can be designed as

$$\dot{x}_o = \begin{cases} 0, & d \geq d_o, \\ \gamma(d)\boldsymbol{v}_{max}, & d_c < d < d_o, \\ \boldsymbol{v}_{max}, & d \leq d_c, \end{cases} \tag{5.6}$$

where $\gamma(d) = (d_o - d)/(d_o - d_c)$, $d = ||\boldsymbol{p}_{cr} - \boldsymbol{p}_{co}||$ is defined as the distance between the obstacle and the manipulator, $\boldsymbol{v}_{max} = v_{max}(\boldsymbol{p}_{cr} - \boldsymbol{p}_{co})/d$ is the maximum avoiding velocity vector with opposite direction to the obstacle velocity's direction, d_o is the threshold distance to avoid the obstacle and d_c is the minimum allowed distance from manipulator to obstacle.

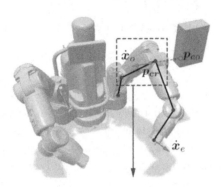

Figure 5.4 Decision of a tentative avoiding velocity $\dot{x}_o$.

(3) Dimension reduction method

If the collision point is found close to the base of the manipulator, the manipulator does not have enough degrees of freedom to achieve the avoidance velocity $\dot{x}_o$. The row rank of the Jacobian matrix J_o might be less than the dimension of $\dot{x}_o$, so it will cause the inverse kinematics problem of the manipulator to be over-defined and the motion of manipulator will be unstable [4].

A dimension reduction method is proposed to solve the aforementioned problem. We know that collision avoidance can be implemented only if the projection of the avoidance velocity equals $\dot{x}_o$, as shown in Fig. 5.5, where N_o is the normal plane of $\dot{x}_o$. Then any velocity vector $\dot{x}'_o$ point on N_o will meet the requirement. So $\dot{x}'_o$ satisfies

$$\dot{x}_o^T(\dot{x}'_o - \dot{x}_o) = 0. \tag{5.7}$$

Then substituting $\dot{x}'_o = J_o\dot{\theta}$ into Eq. (5.7), we can get

$$\dot{x}_o^T\dot{x}_o = \dot{x}_o^T J_o\dot{\theta}. \tag{5.8}$$

When $\dot{x}_o^T\dot{x}_o$ become a scalar, the manipulator will be not over-defined.

(4) Restoration

An artificial parallel system of the manipulator is designed to simulate its position without the influence of the obstacle in real-time. So the manipulator is able to restore its original position when the obstacle has been removed, as shown in Fig. 5.6, where the dashed black line represents an artificial manipulator simulated in the parallel system. This artificial system

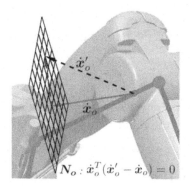

Figure 5.5 Dimension reduction method.

satisfies

$$\dot{\theta}_r = J^{\dagger}(\theta_r)\dot{x}_e, \tag{5.9}$$

where $\dot{\theta}_r$ is the joint velocities of the parallel system.

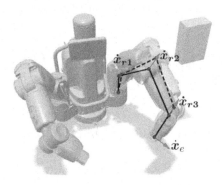

Figure 5.6 Simulated parallel system.

In addition, the dimension reduction is also applied in the parallel system to avoid the over-defined problem, so we have

$$\begin{bmatrix} \dot{x}_{r1} \\ \dot{x}_{r2} \\ \dot{x}_{r3} \end{bmatrix}^T \begin{bmatrix} \dot{x}_{r1} \\ \dot{x}_{r2} \\ \dot{x}_{r3} \end{bmatrix} = \begin{bmatrix} \dot{x}_{r1} \\ \dot{x}_{r2} \\ \dot{x}_{r3} \end{bmatrix}^T \begin{bmatrix} J_{r1} \\ J_{r2} \\ J_{r3} \end{bmatrix} \dot{\theta}, \tag{5.10}$$

where J_{r1}, J_{r2}, and J_{r3} are the Jacobian matrices of the three joints and $\dot{x}_{r1}$, $\dot{x}_{r2}$, and $\dot{x}_{r3}$ are the desired joint velocities moving toward the corresponding joints on the parallel system. The manipulator should satisfy Eq. (5.1) and Eq. (5.10).

We can define $\dot{x} = [\dot{x}_{r1} \quad \dot{x}_{r2} \quad \dot{x}_{r3}]^T$ and $J_r = [J_{r1} \quad J_{r2} \quad J_{r3}]^T$. A closed-loop system is designed for the restoring velocity $\dot{x}_r$ as follows:

$$\dot{x}_r = K_r e_r, \tag{5.11}$$

where K_r is a symmetric positive definite matrix and $e_r = [e_{r1} \quad e_{r2} \quad e_{r3}]^T$ is the position errors of the joints between the parallel system and the real system.

(5) Control design at the kinematic level

Substituting the general solution (5.3) into Eq. (5.8) and Eq. (5.10), we have

$$\dot{x}_o^T J_o J_e^\dagger \dot{x}_e + \dot{x}_o^T J_o (I - J_e^\dagger J_e) z_o = \dot{x}_o^T \dot{x}_o,$$
$$\dot{x}_r^T J_r J_e^\dagger \dot{x}_e + \dot{x}_r^T J_r (I - J_e^\dagger J_e) z_r = \dot{x}_r^T \dot{x}_r. \tag{5.12}$$

The homogeneous solution can be calculated, so we can obtain z_o and z_r as follows:

$$z_o = [\dot{x}_o^T J_o (I - J_e^\dagger J_e)]^\dagger (\dot{x}_o^T \dot{x}_o - \dot{x}_o^T J_o J_e^\dagger \dot{x}_e),$$
$$z_r = [\dot{x}_r^T J_r (I - J_e^\dagger J_e)]^\dagger (\dot{x}_r^T \dot{x}_r - \dot{x}_r^T J_r J_e^\dagger \dot{x}_e). \tag{5.13}$$

In order to avoid an abrupt switch between these two cases and smooth the control, a weighted sum of Eq. (5.13) is used as follows:

$$\dot{\theta}_d = J_e^\dagger \dot{x}_e + (I - J_e^\dagger J_e)[\alpha z_o + (1 - \alpha) z_r], \tag{5.14}$$

where the weight α depends on the distance between the obstacle and the manipulator as

$$\alpha = \begin{cases} 0, & d \geq d_o, \\ \frac{d_o - d}{d_o - d_r}, & d_r < d < d_o, \\ 1, & d \leq d_r, \end{cases} \tag{5.15}$$

where d_r is the threshold of the distance where the manipulator starts to restore its original position.

In conclusion, the control strategy at the kinematic level can be obtained by substituting Eq. (5.2) into Eq. (5.14):

$$\dot{\theta}_d = J_e^\dagger (\dot{x}_d + K_e e_x) + (I - J_e^\dagger J_e)[\alpha z_o + (1 - \alpha) z_r]. \tag{5.16}$$

5.2.2 Dynamics level

(1) RBFNN

Because the RBFNN is able to approximate any continuous function $\phi(\theta)$: $R^m \rightarrow R$ arbitrarily close on a compact set $\Omega_z \subset R^m$ [13], we can use it as follows:

$$\phi(\theta) = \boldsymbol{W}^T \boldsymbol{Z}(\theta) + \varepsilon_\phi, \ \forall \theta \subset \Omega_\theta, \tag{5.17}$$

where $\boldsymbol{W}$ is the ideal NN weight vector of constant numbers, θ is the input vector, ε_ϕ is the bounded approximation error, and $\boldsymbol{Z}(\theta) = [z_1(\theta), z_2(\theta), ..., z_l(\theta))]^T$ is the basis function. In additional, the element of $\boldsymbol{Z}(\theta)$, $z_i(\theta)$, is chosen as Gaussian functions as follows [14,15]:

$$z_i(\theta) = exp\left[\frac{-(\theta - \mu_i)^T(\theta - \mu_i)}{\eta_i^2}\right], \ i = 1, 2, ..., l, \tag{5.18}$$

where μ_i denotes the note centers and η_i denotes the variance. In order to minimize the approximation error for all θ, $\boldsymbol{W}$ is defined as follows:

$$\boldsymbol{W} = \arg\min_{W' \in R^l}\{sup|\phi(\theta) - \boldsymbol{W'}^T \boldsymbol{S}(\theta)|\}. \tag{5.19}$$

When the number of NNs l is large enough and node centers μ_i are approximately chosen, the approximation error can be made arbitrarily small.

(2) Error transformation and joint position control loop

The joint angle error can be defined as $e_\theta = \theta - \theta_d$. According to Refs. [16, 17], we can employ the error transformation function as follows:

$$e_{\theta i}(t) = \rho(t)R_i\left(P_i\left(\frac{e_{\theta i}(t)}{\rho(t)}\right)\right), \ i = 1, 2, ..., n, \tag{5.20}$$

where

$$R_i(x) = \begin{cases} \frac{e^x - \delta}{1 + e^x}, & e_{\theta i}(0) \geq 0, \\ \frac{\delta e^x - 1}{1 + e^x}, & e_{\theta i}(0) < 0. \end{cases} \tag{5.21}$$

In addition, the inverse function of $R_i(x)$ is shown as follows:

$$P_i(x) = \begin{cases} ln\frac{x+\delta}{1-x}, & e_{\theta i}(0) \geq 0, \\ ln\frac{x+1}{\delta-x}, & e_{\theta i}(0) < 0. \end{cases} \tag{5.22}$$

Moreover, we define the tracking performance requirement $\rho(t)$ as follows:

$$\rho(t) = (\rho_0 - \rho_\infty)e^{-pt} + \rho_\infty, \qquad (5.23)$$

where δ, ρ_0, ρ_∞, and p are positive and chosen by the designer.

The joint position control loop aims to generate the desired joint angular velocities and to guarantee the transient performance of the manipulator, so it is designed as follows:

$$v_{id}(t) = -k_1 \rho(t)\eta_i(t) + \dot{\theta}_{di}(t) + \frac{\dot{\rho}(t)}{\rho(t)}e_{\theta i}(t), \qquad (5.24)$$

where

$$\eta_i(t) = P_i\left(\frac{e_{\theta i}(t)}{\rho(t)}\right). \qquad (5.25)$$

We can prove that if $\eta_i(t)$ is bounded, the tracking performance can be tailored by the performance requirement function $\rho(t)$ [16,17]; $\rho(t)$ is able to regulate both the transient and steady-state performance.

Then we can consider a Lyapunov function $V_1 = \frac{1}{2}\eta^T(t)\eta(t)$, which is used to analyze the stability later, and its derivative can be calculated as follows:

$$\dot{V}_1 = \frac{\eta^T(t)\dot{P}(\eta(t))e_v(t)}{\rho(t)} - k_1\eta^T(t)\dot{P}(\eta(t))\eta(t), \qquad (5.26)$$

where $\dot{P}(\eta(t)) = \mathrm{diag}(\dot{P}_1(R_1(\eta(t))), ..., \dot{P}_n(R_n(\eta(t))))$, $v_d = [v_{d1}, v_{d2}, ..., v_{dn}]^T$, and $e_v = \dot{\theta} - v_d$.

(3) Neural learning and joint velocity control loop

The dynamics of the manipulator can be described as

$$M(\theta)\ddot{\theta} + C(\theta, \dot{\theta})\dot{\theta} + G(\theta) = \tau, \qquad (5.27)$$

where $M(\theta)$ is the inertia matrix of the manipulator, $C(\theta, \dot{\theta})$ is the Coriolis matrix of the manipulator, τ_{ext} is the external torque generated by payload, and $G(\theta) = G'(\theta) + \tau_{ext}$, where $G'(\theta)$ is the gravity term.

The velocity control loop aims to achieve the desired joint angular velocities v_d by regulating the control torque τ, which can be designed as follows:

$$\tau = -k_2 e_v + \hat{M}\dot{v}_d + \hat{C}v_d + \hat{G} + \hat{f} - \frac{\dot{P}(\eta(t)\eta(t))}{\rho(t)}, \qquad (5.28)$$

where the variables with a hat denote the estimates and f will be introduced later.

Moreover, the closed-loop dynamics can be formulated as follows:

$$M\dot{e}_v + Ce_v + k_2 e_v + \frac{\dot{P}(\eta(t)\eta(t))}{\rho(t)} - \hat{f} = -(M - \hat{M})\dot{v}_d - (C - \hat{C})v_d - (G - \hat{G}).$$

(5.29)

The NN approximation technique can be applied:

$$\begin{aligned}
M(\theta) &= W_M^T Z_M(\theta) + \varepsilon_M(\theta), \\
C(\theta, \dot{\theta}) &= W_C^T Z_C(\theta, \dot{\theta}) + \varepsilon_C(\theta, \dot{\theta}), \\
G(\theta) &= W_G^T Z_G(\theta) + \varepsilon_G(\theta), \\
f &= W_f^T Z_f(\theta, \dot{\theta}, v_d, \dot{v}_d) + \varepsilon_f(\theta, \dot{\theta}, v_d, \dot{v}_d),
\end{aligned}$$

(5.30)

where W_M, W_C, W_G, and W_f are the ideal NN weight matrices, where

$$\begin{aligned}
W_M &= [W_{M_{i,j}}], \\
W_C &= [W_{C_{i,j}}], \\
W_G &= diag(W_{G_i}), \\
W_f &= diag(W_{f_i}),
\end{aligned}$$

(5.31)

where all elements $\in R^l$ and they are defined in Eq. (5.19).

The matrices of RBFs $Z_M(\theta)$, $Z_C(\theta, \dot{\theta})$, $Z_G(\theta)$, and $Z_f(\theta)$ are designed as follows:

$$\begin{aligned}
Z_M(\theta) &= diag(Z_\theta, ..., Z_\theta), \\
Z_C(\theta, \dot{\theta}) &= diag\left(\begin{bmatrix} Z_\theta \\ Z_{\dot{\theta}} \end{bmatrix}^T, ..., \begin{bmatrix} Z_\theta \\ Z_{\dot{\theta}} \end{bmatrix}^T \right), \\
Z_G(\theta) &= [Z_\theta^T, ..., Z_\theta^T]^T, \\
Z_f(\theta, \dot{\theta}, v_d, \dot{v}_d) &= [\bar{Z}^T, ..., \bar{Z}^T]^T,
\end{aligned}$$

(5.32)

where $Z_\theta = [z_1(\theta), z_2(\theta), ..., z_l(\theta)]^T \in R^l$, $Z_{\dot{\theta}} = [z_1(\dot{\theta}), z_2(\dot{\theta}), ..., z_l(\dot{\theta})]^T \in R^l$, and $\bar{Z} = [Z_\theta^T, Z_{\dot{\theta}}^T, Z_{v_d}^T, Z_{\dot{v}_d}^T]^T \in R^{4l}$.

As for the elements of $\bar{Z}$, $Z_{v_d} = [z_1(v_d), z_2(v_d), ..., z_l(v_d)]^T \in R^l$, $Z_{\dot{v}_d} = [z_1(\dot{v}_d), z_2(\dot{v}_d), ..., z_l(\dot{v}_d)]^T \in R^l$, where z_i is defined in Eq. (5.18).

Moreover, f is defined as follows:

$$f = \varepsilon_M \dot{v}_d + \varepsilon_C v_d + \varepsilon_G.$$

(5.33)

Therefore, the NN-based estimates can be formulated as follows:

$$
\begin{aligned}
\hat{M}(\theta) &= \hat{W}_M^T Z_M(\theta), \\
\hat{C}(\theta, \dot{\theta}) &= \hat{W}_C^T Z_C(\theta, \dot{\theta}), \\
\hat{G}(\theta) &= \hat{W}_G^T Z_G(\theta), \\
\hat{f} &= \hat{W}_f^T Z_f(\theta, \dot{\theta}, v_d, \dot{v}_d).
\end{aligned}
\tag{5.34}
$$

Then, we can substitute Eq. (5.34) into Eq. (5.29). We obtain

$$
\begin{aligned}
M\dot{e}_v + Ce_v + k_2 e_v + \frac{\dot{P}(\eta(t)\eta(t))}{\rho(t)} = &-\tilde{W}_M^T Z_M \dot{v}_d - \tilde{W}_C^T Z_C v_d \\
&- \tilde{W}_G^T Z_G - \tilde{W}_f^T Z_f - \varepsilon_f,
\end{aligned}
\tag{5.35}
$$

where $\tilde{W}_{(\cdot)} = W_{(\cdot)} - \hat{W}_{(\cdot)}$.

Moreover, we can consider the second Lyapunov function $V_2 = \frac{1}{2}e_v^T M_{e_v} + \frac{1}{2}tr(\tilde{W}_M^T Q_M \tilde{W}_M) + \frac{1}{2}tr(\tilde{W}_C^T Q_C \tilde{W}_C + \tilde{W}_G^T Q_G \tilde{W}_G + \tilde{W}_f^T Q_f \tilde{W}_f)$, where Q_M, Q_C, Q_G, and Q_f are positive matrices chosen by the designer.

Note that $\dot{\tilde{W}}_{(\cdot)} = -\dot{\hat{W}}_{(\cdot)}$, and the derivative of V_2 is

$$
\begin{aligned}
\dot{V}_2 = &-e_v^T k_2 e_v - e_v^T \varepsilon_f - \frac{e_v^T \dot{P}(\eta(t))\eta(t)}{\rho(t)} \\
&- tr\left[\tilde{W}_M^T \left(Z_M \dot{v}_d e_v^T + Q_M \dot{\hat{W}}_M\right)\right] \\
&- tr\left[\tilde{W}_C^T \left(Z_C v_d e_v^T + Q_C \dot{\hat{W}}_C\right)\right] \\
&- tr\left[\tilde{W}_G^T \left(Z_G e_v^T + Q_G \dot{\hat{W}}_G\right)\right] \\
&- tr\left[\tilde{W}_f^T \left(Z_f e_v^T + Q_f \dot{\hat{W}}_f\right)\right].
\end{aligned}
\tag{5.36}
$$

Thus, the neural learning law is formulated as follows:

$$
\begin{aligned}
\dot{\hat{W}}_M &= -Q_M^{-1}\left(Z_M \dot{v}_d e_v^T + \sigma_M \hat{W}_M\right), \\
\dot{\hat{W}}_C &= -Q_C^{-1}\left(Z_C \dot{v}_d e_v^T + \sigma_C \hat{W}_C\right), \\
\dot{\hat{W}}_G &= -Q_G^{-1}\left(Z_G e_v^T + \sigma_G \hat{W}_G\right), \\
\dot{\hat{W}}_f &= -Q_f^{-1}\left(Z_f e_v^T + \sigma_f \hat{W}_f\right),
\end{aligned}
\tag{5.37}
$$

where σ_M, σ_C, σ_G, and σ_f are positive parameters chosen by the designer.

In the following, the boundedness of $\eta(t)$ is established so that both transient and steady-state performance of the manipulator can be guaranteed. An overall Lyapunov function $V = V_1 + V_2$ can be considered, and the derivative of V is

$$
\begin{aligned}
\dot{V} = &-k_1 \eta^T(t)\dot{P}((\eta(t)))\eta(t) - e_v^T k_2 e_v - e_v^T \varepsilon_f \\
&+ tr[\sigma_M \tilde{W}_M^T \hat{M}_M] \\
&+ tr[\sigma_C \tilde{W}_C^T \hat{M}_C] \\
&+ tr[\sigma_G \tilde{W}_G^T \hat{M}_G] \\
&+ tr[\sigma_f \tilde{W}_f^T \hat{M}_f].
\end{aligned}
\tag{5.38}
$$

Considering the definition of $\dot{P}(\eta(t))$ and the inequality obtained by Young's inequality, we have

$$
tr[\tilde{W}_{(\cdot)}^T] \le -\frac{1}{2}\|\tilde{W}_{(\cdot)}\|_F^2 + \frac{1}{2}\|W_{(\cdot)}\|_F^2.
\tag{5.39}
$$

So Eq. (5.38) can be further derived as follows:

$$
\begin{aligned}
\dot{V} \le &-\frac{2k_1}{1+\delta}\|\eta(t)\|^2 - \left(k_2 - \frac{1}{2}\right)\|e_v\|^2 + \varrho \\
&-\frac{\sigma_M}{2}\|\tilde{W}_M\|_F^2 \\
&-\frac{\sigma_C}{2}\|\tilde{W}_C\|_F^2 \\
&-\frac{\sigma_G}{2}\|\tilde{W}_G\|_F^2 \\
&-\frac{\sigma_f}{2}\|\tilde{W}_f\|_F^2,
\end{aligned}
\tag{5.40}
$$

where

$$
\varrho = \frac{\sigma_M}{2}\|W_M\|_F^2 + \frac{\sigma_C}{2}\|W_C\|_F^2 + \frac{\sigma_G}{2}\|W_G\|_F^2 + \frac{\sigma_f}{2}\|W_f\|_F^2 + \frac{1}{2}\epsilon_f^2,
\tag{5.41}
$$

where ϵ_f is the upper limit of $\|\varepsilon_f\|$ over Ω.

It is clear that when $\tilde{W}_M$, $\tilde{W}_C$, $\tilde{W}_G$, $\tilde{W}_f$, $\eta(t)$, and e_v satisfy

$$
\begin{aligned}
&\frac{\sigma_M\|\tilde{W}_M\|_F^2 + \sigma_C\|\tilde{W}_C\|_F^2 + \sigma_G\|\tilde{W}_G\|_F^2 + \sigma_f\|\tilde{W}_f\|_F^2}{2} \\
&+ \frac{2k_1}{1+\delta}\|\eta(t)\|^2 + \left(k_2 - \frac{1}{2}\right)\|e_v\|^2 \ge \varrho,
\end{aligned}
\tag{5.42}
$$

we have $\dot{V} \le 0$.

5.3. EEG-based shared control

Nowadays the relationship between robots and humans becomes increasingly intimate, and better and better human–robot collaboration systems are developed. However, convenient operation of robots is still challenging because of the low sensing ability. A brain–computer interface (BCI) was designed to solve this problem. Based on the operator's EEG signals, robot manipulators can be operated much more easily.

In this subsection, a steady-state visual evoked potential (SSVEP)-based BCI is employed to operate the Baxter robot (Fig. 5.7). In order to make the interactive interface more user-friendly, a visual scheme is proposed in which the video of the robot's vision and stimulating flicking diamonds are fused. Moreover, simulating diamonds with different flicking frequencies are attached to different target objects in the robot's vision video. In order to make operators only need to concentrate on the target object on a monitor via EEG signals, a shared control strategy is proposed where the robot's manipulator can pick up the object and avoid collision automatically.

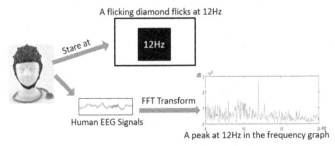

Fig a: The Fundamentals of SSVEP Paradigms

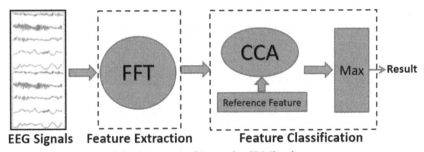

Fig b: The Process of Processing EEG Signals

Figure 5.7 Illustration of BCI.

The proposed system is shown in Fig. 5.8. It can be divided into three parts: BCI, visual servoing (VS), and robot control (RC). The BCI part is designed to collect human EEG signals, which allow it to infer the selected target object. The outcome of the selection is sent to the RC part. The VS part is designed to capture visual feedback from the Baxter robot and the target objects. The target objects in the video are first fused with the flicking diamonds, which are designed to invoke the desired SSVEP EEG signals. Then the fusion video is shown on a monitor that is observed by the user to choose the target objects. Simultaneously, the VS part detects the positions of all target objects and sends them to the robot controller in real-time. The RC part is designed to control the movements of the manipulator of the Baxter robot, including path planning and obstacle avoidance.

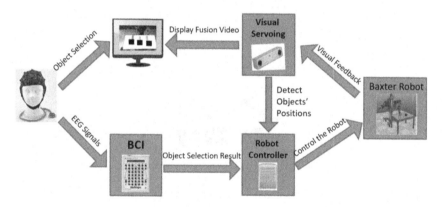

Figure 5.8 Overview of the system.

5.3.1 Visual fusion

(1) Target object detection and display

In this project, an adaptive algorithm for target object detection is proposed based on the colors of the objects. Firstly, the researchers draw the outer contours of the objects. Secondly, the computer calculates the average values of the color of all pixel points inside each contour. The average values represent the features of the objects that can be called "standard color marks." Thirdly, variances between the standard color marks and the color information at every point in the real-time image in experiments are calculated. We consider that the minimum variance value represents the area that is most likely to contain the specific object. Finally, if we succeed in detecting the position of the object, the average color value of the area around the center of the detected object will be calculated, and this is the new value to update the standard color mark. Then, we can go back to the beginning.

After we have detected the positions of the objects in the image, the SSVEP flicking diamonds are attached to each object. In order to ensure the stability of the flicking frequencies, Microsoft DirectX software is used to control the rate of the diamonds which turn white or black. Hence, we can guarantee the accurate flicking frequency of each diamond will evoke precise EEG signals. The main process is shown in Fig. 5.9.

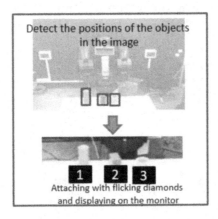

Figure 5.9 The process of object detection and display.

(2) Binocular vision and depth measurement

In this experiment, we use a Bumblebee2 camera. Because it is a binocular camera, the positions of the same objects are different in the left-eye image and the right-eye image. The difference between two images can be calculated by the following equation:

$$d = x_{left} - x_{right}, \tag{5.43}$$

where x_{left} and x_{right} are the x-coordinates of the same object in the left-eye image and the right-eye image, respectively.

The x-, y-, and z-coordinates of the objects in the camera coordinate system can be computed by the x-, y-, and z-coordinates of the objects in the image coordinate system as follows:

$$z_c = \frac{f * L}{f},$$

$$x_c = \frac{z_c(x_{left} - X_{image} * 0.5)}{f}, \tag{5.44}$$

$$y_c = \frac{z_c(y_{left} - Y_{image} * 0.5)}{f},$$

where f is the focal distance of the camera and L is the distance between two sensors of Bumblebee2. The constants X_{image} and Y_{image} are the width and height of the image, respectively. By Eqs. (5.43)–(5.44), the 3D coordinates of all objects can be computed.

(3) Coordinate calibration and improvement

After we have obtained the 3D coordinates of the objects in the camera coordinate system, we should transform the objects' coordinates from the camera coordinate system to the robot coordinate system. We can compute this as follows:

$$T * \begin{bmatrix} x_c \\ y_c \\ z_c \\ 1 \end{bmatrix} = \begin{bmatrix} x_r \\ y_r \\ z_r \\ 1 \end{bmatrix}, \tag{5.45}$$

where $T \in 4 \times 4$ is the transformation matrix that we will obtain by calibration, T_{ci} is the ith column of the T matrix, $[x_c, y_c, z_c, 1]^T$ are the coordinates in the Bumblebee2 coordinate system, and $[x_r, y_r, z_r, 1]^T$ are the coordinates in the Baxter robot coordinate system.

In order to reduce the errors caused by the imprecision of the measurements in the robot's Application Programming Interfaces (APIs) and in the Bumblebee2 camera, the least squares method (LSM) is used for calibration.

The LSM formula is

$$A * P = B, \tag{5.46}$$

where $A \in m \times n$ and $B \in m \times 1$ are matrices with constant values and $P \in n \times 1$ denotes the unknown parameters to be solved.

We used the LSM to find a solution to P, which could identify the least squares errors. The vector of P was calculated according to the LSM theory as follows:

$$P = (A^T * A)^{-1} * A^T * B. \tag{5.47}$$

In this study, we proposed a transformation method to employ the LSM algorithm in the calibration. After a series of mathematical deductions, Eq. (5.45) can be transformed into the format of the following equation to fit the expression of the LSM method:

$$
\begin{bmatrix}
x_{c1}I_4 & y_{c1}I_4 & z_{c1}I_4 & I_4 \\
x_{c2}I_4 & y_{c2}I_4 & z_{c2}I_4 & I_4 \\
\vdots & \vdots & \vdots & \vdots \\
x_{cn}I_4 & y_{cn}I_4 & z_{cn}I_4 & I_4
\end{bmatrix}
\begin{bmatrix}
T_{c1} \\
T_{c2} \\
T_{c3} \\
T_{c4}
\end{bmatrix}
=
\begin{bmatrix}
x_{r1} \\
y_{r1} \\
z_{r1} \\
1 \\
\vdots \\
x_{rm} \\
y_{rm} \\
z_{rm} \\
1
\end{bmatrix},
\tag{5.48}
$$

where $I_4 \in R^{4\times4}$ is an identity matrix, T_{ci} is the ith column vector in the transformation matrix T as in Eq. (5.45), and $[x_{ci}, y_{ci}, z_{ci}]$ and $[x_{ri}, y_{ri}, z_{ri}]$ are the coordinates of a series of points measured in the Bumblebee2 coordinate system and the Baxter coordinate system, respectively.

We can express Eq. (5.48) as follows:

$$
X_{cLSM} * T_{LSM} = X_{rLSM}.
\tag{5.49}
$$

The vector T_{LSM} in this equation can be solved as follows:

$$
T_{LSM} = (X_{cLSM}^T * X_{cLSM})^{-1} * X_{cLSM}^T * X_{rLSM}.
\tag{5.50}
$$

After we calculated the value of the vector T_{LSM}, we transform it into the form of the transform matrix T according to Eq. (5.45), which is 4×4. With this process, the calibration problem could be solved by the LSM.

5.3.2 Mind control

(1) The BCI paradigm of SSVEP

Because SSVEP has the highest response and requires less training than the others, it is a relatively suitable paradigm for this robotic system. The BCI paradigm of SSVEP is shown in Fig. 5.7(a). In order to get EEG signals from the operators, the widely used Neuroscan amplifier device with 40 channels is used.

In this paradigm of the SSVEP BCI, different diamonds with different flicking frequencies (such as 10 Hz, 12 Hz, and 15 Hz) are attached to different objects and the fused video is displayed on the screen. If the operator considers one of the objects as the target, he or she should stare at the object, and undoubtedly he or she is staring at the specific flicking diamond at the same time. So we can collect the EEG signals, process them, and infer the target which the operator wants.

(2) EEG signal processing

The process of analyzing EEG signals is shown in Fig. 5.7(b). Firstly, the fast Fourier transform (FFT) is used to extract the frequency features.

Secondly, the canonical correlation analysis (CCA) algorithm is employed to classify the EEG signals. The main process of CCA is as follows [18].

Considering two variables $S_x = X_1, X_2, ..., X_n$ (where $X_i \in m \times 1$) and $S_y = Y_1, Y_2, ..., Y_n$ (where $Y_i \in m \times 1$), we first perform the linear transformation of the two variables as follows:

$$
\begin{aligned}
S_x \omega_x &= (\omega_{x1} X_1, \omega_{x2} X_2, ..., \omega_{xn} X_N), \\
S_y \omega_y &= (\omega_{y1} Y_1, \omega_{y2} Y_2, ..., \omega_{yn} Y_N),
\end{aligned}
\tag{5.51}
$$

where ω_x and ω_y are linear transformation matrices.

CAA is applied to determine the best transformation matrices that maximize the relationship between S_x and S_y. This relationship is estimated by the following equation:

$$
\rho = \max coor(S_x \omega_x, S_y \omega_y) = \max \frac{< S_x \omega_x, S_y \omega_y >}{||S_x \omega_x|| ||S_y \omega_y||}.
\tag{5.52}
$$

Through further derivation using the techniques reported in Ref. [18], the maximal solution of Eq. (5.52) is

$$
\omega_y = \frac{C_{yy}^{-1} C_{yx} \omega_x}{\lambda},
\tag{5.53}
$$

$$
C_{xy} C_{yy}^{-1} C_{yx} \omega_x = \lambda^2 C_{xx} \omega_x,
\tag{5.54}
$$

where C_{xx} and C_{yy} are covariance matrices. It is symmetrically and positively definite; therefore, it can be decomposed into two matrices:

$$
C_{xx} = R_{xx} R'_{xx}.
\tag{5.55}
$$

We define

$$u_x = R'_{xx}\omega_x. \tag{5.56}$$

Then, Eq. (5.54) can be expressed as

$$R_{xx}^{-1} C_{xx} C_{yy}^{-1} C_{yz} R_{xx}^{-1'} u_x = \lambda^2 u_x. \tag{5.57}$$

We define $A = R_{xx}^{-1} C_{xx} C_{yy}^{-1} C_{yz} R_{xx}^{-1'}$. Then Eq. (5.57) becomes $Au_x = \lambda^2 u_x$. Therefore, u_x can be calculated as the feature vector of matrix A. Based on Eq. (5.56) and Eq. (5.53), ω_x and ω_y can then be computed. Then the maximal relationship between two variables can be computed by Eq. (5.52).

By employing the algorithm described above, we established the correlation between the EEG signal and the reference signals. By comparing them and taking the maximum value, we obtained the case that the user would be the most likely to select in his or her mind.

(3) Kinematics of the Baxter robot and obstacle avoidance

When the coordinates of the target object have been sent to the robotic controller, the end-effector of the manipulator will autonomously arrive at the destination to pick up the object. Simultaneously, the controller will detect all obstacles around the manipulator and plan an appropriate trajectory to avoid collisions with obstacles in the environment.

According to the model of the Baxter robot [19], the coordinate of the end-effector in Cartesian space x and the joint angle vector q of the Baxter robot manipulator satisfies the following equation:

$$x = f(q), \tag{5.58}$$

where $x \in R^6$, $q \in R^7$, and f is a nonlinear mapping from the joint angle space to Cartesian space [19].

After differentiating Eq. (5.58), we can obtain the linear equation as follows:

$$\dot{x} = J\dot{q}, \tag{5.59}$$

where $\dot{x} \in R^6$ is the velocity of the end-effector in Cartesian space, $\dot{q} \in R^7$ is the angular velocity vector consisting of all the angular velocity values of every joint in the manipulator, and $J \in R^{6 \times 7}$ is the Jacobian matrix.

Because the manipulators are redundant, we can divide the solution to Eq. (5.59) into two parts:

$$\dot{q} = \dot{q}_t + \dot{q}_s, \tag{5.60}$$

where $\dot{q}_t$ is the component that performs the task motion and the particular solution of Eq. (5.59) and $\dot{q}_s$ is the performance of self-motion and the homogeneous solution to Eq. (5.59) whose value does not affect the position of the end-effector (the value of x); it changes only the position of the manipulator. Therefore, by adjusting the value of $\dot{q}_s$, we can ensure that the manipulator avoids obstacles by proper positions without influencing the task performance of $\dot{q}_t$.

In this experiment, we proposed a method for the coordination of task motion and self-motion (CTS), which can make the manipulator avoid potential obstacles [20].

First, an indicator was defined to estimate the distance between the obstacles and the manipulator:

$$H = d_{min} = min(d_i), \ i = 1, 2, ..., 7, \tag{5.61}$$

where d_i is the distance between the obstacles and the link i of the manipulator and H is the minimum distance between d_i and the distance indicator.

Second, aiming to plan the motions of the manipulator in real-time, the derivative of Eq. (5.61) is calculated as follows:

$$\Gamma = \frac{dH}{dt} = \frac{\partial H}{\partial q}\frac{dq}{dt} = \nabla H \cdot \dot{q}. \tag{5.62}$$

It is obvious that the value of the differential of H (i.e., the value of Γ) can be used to estimate the effect of the manipulator's motions.

Then, according to Eq. (5.60), we can further divide Γ into two parts (Γ_t and Γ_s):

$$\Gamma_t = \nabla H \cdot \dot{q}_t,$$
$$\Gamma_s = \nabla H \cdot \dot{q}_s, \tag{5.63}$$
$$\Gamma = \nabla H \cdot \dot{q} = \nabla H \cdot \dot{q}_t + \nabla H \cdot \dot{q}_s = \Gamma_t + \Gamma_s,$$

where Γ_t and Γ_s are used to estimate the effects of $\dot{q}_t$ and $\dot{q}_s$, which are similar to Γ. If $\Gamma_t > 0$, the trajectory of the manipulator is safe, and the manipulator is moving away from the obstacles. If $\Gamma_t < 0$ and H are less than a threshold, the manipulator is close to the obstacles and moves toward them. Therefore, a collision is likely to occur. In this case, we start the action of $\dot{q}_s$ to avoid collision.

(4) Generation of the task motion and the self-motion

According to Ref. [20], Eq. (5.60) can be expressed as follows:

$$\dot{q} = G_1\dot{x} + (G_2J - I_n)Z = G_1\dot{x} + k(I_n - G_2J)\nabla H, \tag{5.64}$$

where G_1 and G_2 are generalized inverse matrices of J, that is, $JG_1J = J$, $JG_2J = J$, and I_n is the identity matrix with dimension $n \times n$.

As for $Z \in 7 \times 1$, it is an arbitrary vector which can be set as $Z = -k\nabla H$, where k is a real scalar and ∇H is the gradient of H in Eq. (5.61).

By Eq. (5.64), the values of $\dot{q}_t$ and $\dot{q}_s$ could be generated. Furthermore, by changing the parameters in G_2, we can adjust the value of Γ_s in Eq. (5.63) to be greater than 0 or less than 0 in order to adjust the effect of the self-motion.

The obstacle experiment is shown in Fig. 5.10. In Fig. 5.10(a), the Baxter manipulator is illustrated as linked lines. Comparing Fig. 5.10(c) with Fig. 5.10(b), we see that when the CTS method was used, the manipulator was able to avoid the obstacle.

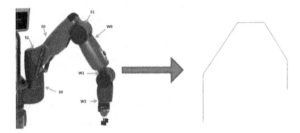

Figure a: Abstract the Manipulator as Several Linked Lines

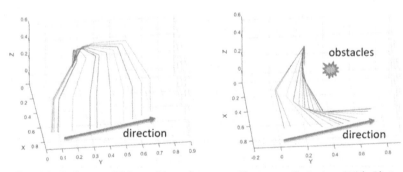

Figure b: Trajectory Without Obstacles   Figure c: Trajectory With Obstacles

Figure 5.10 Results of the obstacle avoidance experiment.

5.4. MR-based user interactive path planning

This subsection proposes a shared control system for the path planning of an omnidirectional mobile robot (OMR) based on MR. Most research on mobile robots is carried out in a completely real environment or a completely virtual environment like using virtual reality (VR). However, with the increasing practical applications of VR, mobile robots sometimes need to avoid some virtual objects, such as virtual meetings that need to avoid projecting participants or factories that need to avoid certain locations. In these scenarios, a real environment containing virtual objects has important actual applications.

The proposed system can control the movement of the mobile robot in the real environment, as well as the interaction between the mobile robot's motion and virtual objects which can be added to a real environment. It mainly consists of two components. Firstly, an MR interaction system was developed to control the mobile robot. Additionally, virtual obstacles can be added to the map of the robot's movement based on specific needs, and path planning can be actively modified. Secondly, the original vector field histogram* (VFH*) algorithm was modified in terms of threshold setting, candidate direction selection, and the cost function, making the generated path more suitable for the scene in this paper, safer, and smoother.

The diagram of this proposed system is shown in Fig. 5.11. It can control the path planning of the mobile robot through the interface of MR and realize interactions between the real environment and virtual objects in an innovative manner. The wearer of the HoloLens uses it to send commands to the computer and receive images. After receiving the commands, the computer combines the corresponding path planning algorithm to generate the optimal path and then sends it to the mobile robot to perform the motion process.

5.4.1 Mixed reality

(1) An overview of mixed reality

VR is a familiar technology, and there are many applications in our lives, such as Oculus Rift [21]. In contrast to VR, which completely immerses the user in the virtual environment, MR is a technology that superimposes the virtual environment onto the real environment and allows the user to interact with the virtual environment [22].

In addition, robot control-based MR can solve problems existing in robot control-based VR, such as transmission delay and inaccuracy of the

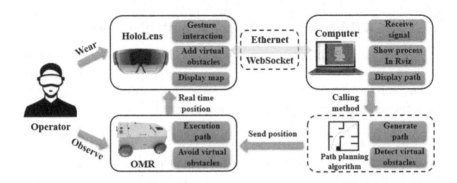

Figure 5.11 The diagram of the system.

virtual robot state. When the operator using an MR device is immersed in the virtual environment, it can also see the real environment. Therefore, the operator can interact with the robot through the virtual environment while looking at the state of the real robot. While giving full play to the advantages of MR, the influence on the control effect of the robot is reduced.

(2) Microsoft HoloLens

HoloLens is a head-mounted MR device with a Windows system installed, as shown in Fig. 5.12. It is equipped with a 32-bit central processing unit (CPU), a graphics processing unit (GPU), and a special holographic processing unit (HPU). Moreover, various information fusions of the inertial measurement unit (IMU), the environment understanding camera, a depth camera, and an ordinary camera enable the device to literally render the environment virtual, giving the wearer a better sense of immersion [23].

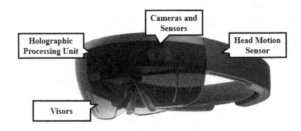

Figure 5.12 The mixed reality device HoloLens.

5.4.2 VFH* algorithm

In this subsection, we introduce the original VFH* algorithm and then propose modifications. We modified the original VFH* algorithm via threshold setting, candidate direction selection, and a cost function to make it more suitable for the scenario of this chapter. When the mobile robot deals with virtual obstacles that are suddenly added to the map, it can have a better obstacle avoidance effect.

(1) The original VFH* algorithm

The vector field histogram+ (VFH+) method determines the forward direction of the robot through the following five steps [24]:

1. Generating the polar histogram: The VFH+ algorithm divides the active region of the current robot position into multiple sectors and calculates the obstacle density in each sector. Then the density of each sector is arranged into a histogram according to the sector number.
2. Binarization polar histogram: One must select the appropriate threshold according to the actual situation and binarize the histogram generated in the previous step. Sectors above the threshold are set as impassable areas, while sectors below the threshold are set as passable areas.
3. Masking the polar histogram: Considering the kinematics and the dynamics characteristics of the robot, the current inaccessible sectors are set as the impassable areas.
4. Determining the direction of motion: The passable areas in the polar histogram is used as the candidate direction; the cost is calculated according to the cost function; and the cost of the passable area is sorted. A commonly used cost function is

$$g(c) = \mu_1 \cdot \Delta(c, k_t) + \mu_2 \cdot \Delta(c, \frac{\theta}{\alpha}) + \mu_3 \cdot \Delta(c, k_p), \qquad (5.65)$$

where c denotes the candidate direction, $g(c)$ is the cost value of the direction, μ_1, μ_2, and μ_3 are three parameters that we need to determine according to the actual situation, $\Delta(c, k_t)$ is the absolute difference between the candidate direction c and the target direction k_t, $\Delta(c, \frac{\theta}{\alpha})$ represent the difference between the candidate direction c and the robot's orientation $\frac{\theta}{\alpha}$, and $\Delta(c, k_p)$ represents the difference between the candidate direction c and the previous direction k_p.

5. The passable area with the lowest cost is selected as the forward direction of the robot.

Since VFH+ only locally plans in real-time, global optimization cannot be guaranteed. Therefore, based on the VFH+ and A* algorithms,

Iwan Ulrich and Johann Borenstein proposed the VFH* algorithm. The A* algorithm is the most effective direct search method to determine the shortest path in static road networks [25]. The VFH* algorithm uses the VFH+ algorithm to predict several future states of the trajectory and form a path tree. Then, the A* algorithm is used to search the path tree to find the global optimal plan as the next movement direction of the robot's current position.

(2) Threshold setting

After generating the polar histogram, the selection of threshold values is very important to determine the direction. If the threshold is too high, the mobile robot will neglect some obstacles on the map, leading to collisions. At the same time, a high threshold will produce more candidate directions and increase the calculation load. If the threshold is too low, the choice of candidate direction will be reduced. When a narrow, passable road section is encountered, the passable direction is neglected by a low threshold, precluding smooth movement.

In order to make the robot adapt better in the changeable environment, a threshold of dynamic change is adopted (Fig. 5.13). When the mean value of polar coordinates in the polar histogram is less than or equal to the specified value, the small threshold value is selected (Fig. 5.13(a)). When it is greater than the specified value, the large threshold is selected (Fig. 5.13(b)).

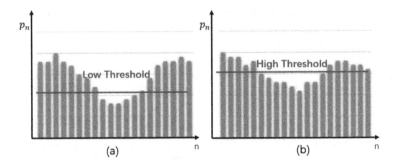

Figure 5.13 Threshold setting.

The change of threshold is related to the mean value of the polar coordinates:

$$T_1 = \begin{cases} t_{low} & (\sum_n p_n \leq \delta_p), \\ t_{high} & (\sum_n p_n > \delta_p), \end{cases} \qquad (5.66)$$

where p_n is the polar histogram of each sector and δ_p is a parameter that distinguishes the size of the density.

However, we should consider the special case where the target point is located between the obstacle and the current position of the robot and the obstacle is within the detectable range. In this case, the direction to the target point is detected as impassable, so the robot cannot reach the target point. To avoid this result, we need to set a special and large threshold for this case to make the direction of the target point passable.

Therefore, the complete process of selecting the threshold is

$$T_2 = max(T_1, \frac{\gamma}{d_{rt}^2}), \qquad (5.67)$$

where d_{rt} is the distance from the mobile robot to the target point and γ is a custom parameter that determines the threshold change condition.

(3) Candidate direction selection

After determining the passable sectors, we add the process of sieving the sectors. The policy of selecting the appropriate direction we proposed is as follows.

1. If the passable sector is the whole circle, which means the robot can move in any direction, the direction towards the target point is undoubtedly a better choice. To avoid unnecessary cost calculation and ensure the accuracy of direction selection, we reduce the passable sector in this case to the area near the target sector (Fig. 5.14(a)).

2. If the passable sector is large (greater than 80 degrees), the direction close to the sector boundary is not a good choice. We prefer choosing some directions inside the sector, which theoretically are safer. Therefore, for the passable sectors, we also made an appropriate reduction (Fig. 5.14(b)).

3. As the obstacles in the constructed map will be inflated, the passage area will be safer. However, when some obstacles are close to each other, the operation of obstacle inflation will make the previously passable area impassable due to the narrow detected passable area. To reduce the occurrence of the above situation, if the detected passable sector angle

is between 10 and 20 degrees, we will adjust the passable angle to make it passable in the above situation (Fig. 5.14(c)).

4. If the range of candidate directions does not fall into the above three cases, the generated candidate directions are not adjusted.

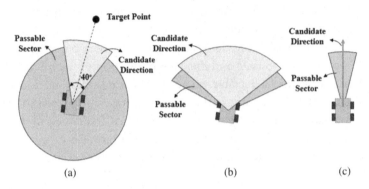

(a)   (b)   (c)

Figure 5.14 Candidate direction selection.

In conclusion, the above candidate direction selection strategy is as follows:

$$C_d = \begin{cases} [c_{rt} - \dfrac{20°}{\alpha}, c_{rt} + \dfrac{20°}{\alpha}] & (\beta = 360°), \\[2mm] [c_l - \dfrac{15°}{\alpha}, c_r + \dfrac{15°}{\alpha}] & (80° < \beta < 360°), \\[2mm] \dfrac{c_l + c_r}{2} & (10° < \beta < 20°), \\[2mm] [c_l, c_r] & \text{otherwise,} \end{cases} \qquad (5.68)$$

where C_d is the range of candidate directions sieved, c_{rt} is the sector from the robot's current position to the target point, β is the angle of the passable sectors, α is the angular resolution of sectors, and c_l and c_r are the left and right boundaries of the passable sector.

(4) The cost function

When there are several candidate directions, their cost values become the key as to which is chosen as the final direction. Considering we use OMR in this experiment, the cost function does not need to take into account the current robot angle, in contrast to Eq. (5.65).

Since the VFH* algorithm uses a path tree to judge the cost, for a generated path tree, the cost of successor nodes should be related to the cost of previous nodes, which is similar to the following A* algorithm heuristic function:

$$f(n) = g(n) + h(n), \tag{5.69}$$

where $g(n)$ is the cost from the starting node to the current node n, $h(n)$ is the estimated cost from the current node n to the target node, and $f(n)$ is the last cost of the current node n.

The cost function of VFH* should also be heuristic and include the cost of previous nodes in the path tree. Therefore, we use the following function:

$$\begin{cases} g(c_n) = \lambda_1 \cdot \Delta(c_n, d_t) + \lambda_2 \cdot \Delta(c_n, d_{n-1}), \\ g(c_{n+i}) = \lambda_3 \cdot g(c_{n+i-1}) + \lambda_4 \cdot \Delta(c_{n+i}, d_t) \\ \qquad + \lambda_5 \cdot \Delta(c_{n+i}, d_{n+i-1}), \end{cases} \tag{5.70}$$

where c_n is the candidate direction from the robot's current position to the next node, c_{n+i} is the candidate direction of the ith node in a path tree when generating the tree, $\Delta(c_n, d_t)$ is the absolute difference of the sector between the candidate direction and the target direction, and $\Delta(c_n, d_{n-1})$ is the absolute difference of the sector between the candidate direction and the previous direction.

When calculating the first node with zero depth in the path tree, the first equation in Eq. (5.70) is used. Only the target node and the previous direction need to be considered. When calculating the successor nodes, the second equation in Eq. (5.70) is used to consider not only the target node and the previous direction, but also the cost of the node before the path tree, and then the cost of this node is synthesized. So, i in Eq. (5.70) should be no more than the depth of the path tree.

For parameter selection, the higher $\lambda_1(\lambda_4)$, the more path selection tends toward the target region; the higher $\lambda_2(\lambda_5)$, the smoother the path selection tends to be. To achieve the effect of shortest path, parameters should meet the following conditions:

$$\lambda_1 > \lambda_2, \; \lambda_4 > \lambda_5. \tag{5.71}$$

At the same time, to make the method more sensitive to sudden virtual obstacles and to avoid obstacles in time to generate a better path, λ_3 parameters should also meet the following conditions:

$$\lambda_3 \cdot \lambda_1 < \lambda_4, \; \lambda_3 \cdot \lambda_2 < \lambda_5. \tag{5.72}$$

After the path tree of a node is generated, the mobile robot will move the distance of a node in the direction with the lowest total cost in the tree and then generate a new path tree at the location of the new node. The above steps are repeated until the robot reaches the target point.

(5) Modification of the VFH* algorithm

Combined with the original process of the VFH* algorithm and the modification proposed in this chapter, the implementation process of the system obstacle avoidance algorithm is shown in Algorithm 1.

Algorithm 1 The improved VFH* algorithm.

Input: Map information, start point, and target point location;
Output: Trajectory;
  1: **while** the distance between the robot and the target point ≤ 0.1 **do**
  2:   Generate polar histogram from the current position;
  3:   The passable candidate direction is screened by combining threshold T_2 and mask C_d;
  4:   The heuristic function $g(c_n)$ and $g(c_{n+i})$ are used to calculate the cost of passable reference direction;
  5:   Choose the direction with the least cost as the final direction;
  6:   Transmit the direction;
  7: **end while**;

5.5. No-target obstacle avoidance with sEMG-based control

In HRI, human biological and psychophysical signals often act as an expression of human intent. In order to capture these human signals, a great number of sensors have been developed. EMG is one of the most widely used physiological signals. Normally, they are collected by surface EMG (sEMG) sensors, such as data gloves and armbands, which are non-invasive and can provide the operator with a better operating experience than intramuscular EMG sensors [26]. As the sEMG signals are highly associated with the activation level of muscles being measured, they can be used for the estimation of muscle stiffness and gesture recognition [27,28].

In this subsection, a novel shared control strategy for mobile robots is proposed. For security reasons, an obstacle avoidance scheme is introduced to the shared control system as collision avoidance guidance. The motion of the mobile robot is a result of compliant motion control and obstacle avoidance. In the mode of compliant motion, the sEMG signals obtained from the operator's forearms are transformed into human commands to control the moving direction and linear velocity of the mobile robot. When the mobile robot is blocked by obstacles, the motion mode is converted into obstacle avoidance. Aimed at the obstacle avoidance problem without a specific target, we develop a no-target Bug (NT-Bug) algorithm to guide the mobile robot to avoid obstacles and return to the command line. Besides, the command moving direction given by the operator is taken into consideration in the obstacle avoidance process to plan a smoother and safer path for the mobile robot. A model predictive controller is exploited to minimize the tracking errors.

The shared control system architecture is shown in Fig. 5.15.

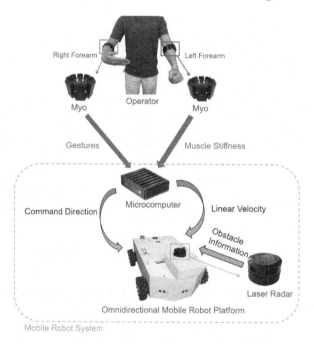

Figure 5.15 The shared control system architecture.

In the control process, the operator is required to wear a MYO Armband on both forearms. The sEMG signals from the left forearm are used to extract muscle stiffness, while those from the right forearm are used for gesture recognition. Then, the muscle stiffness is converted into a linear velocity command for the mobile robot, while the result of gesture recognition is transformed into the moving direction of command motion. The MYO device initially supports recognition of five pre-set gestures, Fist,

`Fingers Spread`, `Wave Left`, `Wave Right`, and `Double Tap`, which are used to command the mobile robot to move forward, backward, left, right, and stop, respectively. We use `command direction` to represent the angle between the direction vector of command motion and the x-axis of the global coordinate system and `command line` to represent the straight line along the `command direction`. The relationship between gestures and `command direction` is shown in Fig. 5.16.

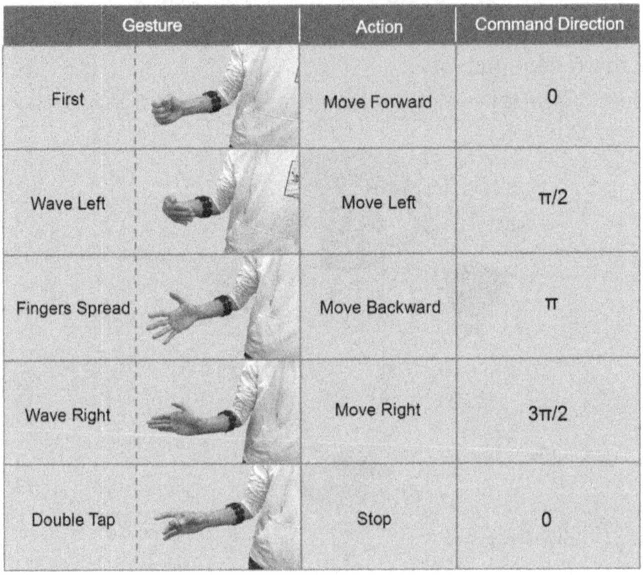

Gesture		Action	Command Direction
First		Move Forward	0
Wave Left		Move Left	$\pi/2$
Fingers Spread		Move Backward	π
Wave Right		Move Right	$3\pi/2$
Double Tap		Stop	0

Figure 5.16 Relationship between gestures and command directions.

5.5.1 Obstacle detection and muscle stiffness extraction

(1) Obstacle detection

In actual applications, we need to be aware of the distance between the mobile robot and obstacles in real-time to take actions to avoid collisions. Based on `Costmap_2d`, we establish a dynamic local cost map centered on the mobile robot using the data from the laser radar to obtain the obstacle information, which is shown in Fig. 5.17.

`Costmap_2d` is an open source algorithm package in the Robot Operating System (ROS), which can automatically produce a 2D cost map based on the data from sensors such as cameras and radars [29]. In the cost map, the environment is divided into a series of grid cells and each of them has a cost value which represents the potential of collisions. Considering the size of the mobile robot, the original cost map is usually processed by an inflation algorithm, which enlarges the influence area of obstacles.

Since the grid cells of obstacles have the largest cost value, we can seek out the grid cells that belong to obstacles according to the cost value and obtain their coordinates relative to the mobile robot. Through calculating the Euclidean distance between the robot center and the grid cells of obstacles, we can obtain the minimum Euclidean distance as the closest obstacle distance.

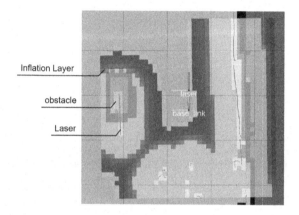

Figure 5.17 Dynamic local cost map.

(2) Muscle stiffness extraction

The sEMG signals from the left forearm of the operator are used to estimate muscle stiffness. As a MYO has eight EMG sensors, it can collect sEMG signals from eight channels at a time. Adding the amplitude of the sEMG signals from eight channels, we have

$$A_{emg}(k) = \sum_{i=1}^{N} | E_i(k) |, \qquad (5.73)$$

where $N = 8$ represents the number of channels, k represents the current sampling instant, and E_i is the sEMG signal from the ith channel. Then, a moving average filter is applied to acquire the sEMG envelop at the sampling instant k:

$$\epsilon_{emg}(k) = \frac{1}{w} \sum_{i=0}^{w} A_{emg}(k-i), \qquad (5.74)$$

where w is defined as the window size. According to Ref. [30], the muscle stiffness can be extracted by

$$\varsigma_{emg}(k) = \frac{e^{\alpha \epsilon_{emg}(k)} - 1}{e^{\alpha} - 1},$$ (5.75)

where e^x is the exponential function and α is a nonlinear shape factor. The extraction of muscle stiffness is shown in Fig. 5.18.

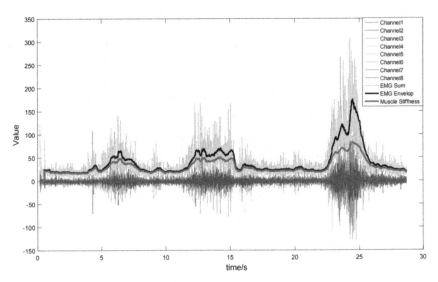

Figure 5.18 Extraction of muscle stiffness ($w = 50$, $\alpha = -0.01$).

(3) Linear velocity

We define the linear velocity of the mobile robot as being proportional to muscle stiffness, that is, as muscle stiffness increases, the linear velocity increases accordingly. Firstly, we normalize the muscle stiffness:

$$\varrho_{emg}(k) = \frac{\varsigma_{emg}(k) - \varsigma_{min}}{\varsigma_{max} - \varsigma_{min}},$$ (5.76)

where $\varrho_{emg}(k)$ can be seen as a control gain and ς_{min} and ς_{max} are the minimum and maximum magnitude of muscle stiffness, respectively.

Then, according to the mapping relationship between muscle stiffness and the linear velocity of the mobile robot, we have

$$v_{linear}(k) = (v_{max} - v_{min})\varrho_{emg}(k) + v_{min},$$ (5.77)

where v_{min} and v_{max} are the minimum and maximum linear velocity of the mobile robot, respectively.

5.5.2 No-target Bug algorithm

Bug is a dynamic path planning method, which is suitable for obstacle avoidance with limited information [31]. It guides the robot to walk along the boundary of obstacles to avoid collisions and takes the original moving direction as a determinant of leaving obstacles.

Inspired by the Bug algorithm, we develop the NT-Bug algorithm, which can guide the mobile robot to avoid both static and dynamic obstacles in the case of no specific target. As the Bug algorithm does not rely on the prior information of the environment, in terms of complicated obstacles, the path generated is inevitably zigzag, which is difficult to track in practice. Therefore, we introduce the moving direction given by the operator into the obstacle avoidance process to plan a smoother and safer path for the mobile robot.

When the mobile robot receives a new moving direction command from the operator, the command line is updated according to the position of the mobile robot and the command direction at the current sampling instant. The command v_x and v_y velocities are the velocity components of linear velocity along the x-axis and the y-axis of the global coordinate system. The function of the NT-Bug algorithm is to guide the mobile robot to avoid obstacles and return to the command line. At every sampling time, the NT-Bug algorithm calculates the moving direction for the mobile robot.

In order to be clear, notations are introduced in Table 5.1.

Table 5.1 Parameters of the no-target Bug algorithm.

O^m	The center of the mobile robot.
B	The closest obstacle cell.
$l(O^m, B)$	The closest obstacle distance.
l_{safe}	The minimum safety distance related to the size of the mobile robot.
θ_{obs}	The direction of the virtual repulsive force exerted by B on the mobile robot.
θ	The moving direction of the mobile robot.
θ_{mode}	The command direction.
θ_{esc}	A once-and-for-all decided upon direction of passing around an obstacle.
d	The distance from O^m to the command line.
d_σ	The distance threshold.
$pattern$	The stage of the NT-Bug algorithm.

Algorithm 2 No-target Bug1.

Input: θ_{esc}, θ_{obs}, ℓ_{obs}, d
Output: θ
  if $\ell_{obs} < \ell_{safe}$ and *pattern* $= 0$ **then**
    pattern $\leftarrow 1$;
    $\theta \leftarrow \theta_{obs} + \theta_{esc}$;
  else if *pattern* $= 1$ **then**
    if d $< d_\sigma$ **then**
      pattern $\leftarrow 0$;
      break;
    else
      $\theta \leftarrow \theta_{obs} + \theta_{esc}$;
    end if
  end if

---

(1) No-target Bug1

The pseudo-code of the NT–Bug1 algorithm is presented in Algorithm 2. Once the closest obstacle distance $\ell(O^m, B)$ is smaller than the minimum safe distance ℓ_{safe}, the motion mode is converted to obstacle avoidance. Inspired by the Bug algorithm, we choose the way of moving around the outline of the obstacles to prevent collisions in the NT–Bug1 algorithm. Therefore, the moving direction of the mobile robot can be obtained by

$$\theta = \theta_{obs} + \theta_{esc}. \tag{5.78}$$

We set *pattern* $= 1$ to represent the stage of moving parallel to the outline of the obstacle. The mobile robot would keep moving in this way until the condition d $< d_\sigma$ is satisfied, which means the mobile robot has already bypassed the obstacle and reached the command line. d_σ is small enough to meet the accuracy requirement of the system. Then, the motion mode is turned to compliant control, and the mobile robot keeps moving along the command line. The procedure of passing a concave obstacle in the NT–Bug1 algorithm is illustrated in Fig. 5.19.

(2) No-target Bug2

The NT–Bug1 algorithm can successfully guide the mobile robot to avoid obstacles and return to the command line. However, as it is not sensitive to the shape of obstacles, the produced path could become tortuous referring to

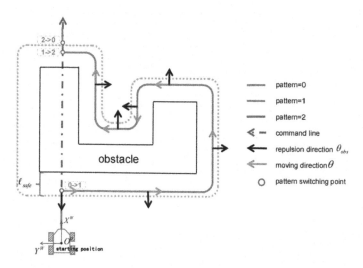

Figure 5.19 The procedure of passing a concave obstacle in the NT-Bug1 algorithm ($\theta_{esc} = \pi/2$).

complicated obstacles, which are difficult to track in practice. Considering the application scenarios are often indoor workplaces where the obstacles are mostly irregularly convex or concave, we introduce the command direction into the obstacle avoidance process to plan a smoother and safer path.

The pseudo-code of the NT-Bug2 algorithm is presented in Algorithm 3 and the procedure of passing a concave obstacle in the NT-Bug2 algorithm is illustrated in Fig. 5.20. As mentioned before, in the stage of *pattern* = 1, the mobile robot moves parallel to the outline of obstacles to prevent collisions.

pattern 1 → 2: The condition $\theta_{obs} = \theta_{mode}$ is satisfied, which means the obstacles could not affect the movement of the mobile robot along the command direction. The mobile robot keeps moving in the current direction to approach the command line.

pattern 1 → 3/2 → 3: The mobile robot reaches the command line and the closest obstacle will not impede its movement, whereas it is still within the distance ℓ_{safe}. It will move in the command direction to be away from the obstacle.

pattern 2 → 1/3 → 1: The mobile robot is blocked by another obstacle. It moves around the outline of the obstacle to avoid it.

pattern 3 → 0: The mobile robot has successfully bypassed the obstacle and moves along the command line. The control mode is converted to compliant control.

Algorithm 3 No–target Bug2.

Input: θ_{mode}, θ_{esc}, θ_{obs}, ℓ_{obs}, d
Output: θ
  if $\ell_{obs} < \ell_{safe}$ and *pattern* = 0 **then**
    pattern ← 1;
  else if *pattern* = 1 **then**
    if $\theta_{obs} = \theta_{mode}$ **then**
      pattern ← 2;
    else if d < d_σ **then**
      pattern ← 3;
    end if
  else if *pattern* = 2 **then**
    if $|\theta_{obs} - (\theta_{mode} + \theta_{esc})| > \frac{\pi}{2}$ **then**
      pattern ← 1;
    else if d < d_σ **then**
      pattern ← 3;
    end if
  else if *pattern* = 3 **then**
    if $\ell_{obs} > \ell_{safe}$ **then**
      pattern ← 0;
      break;
    else if $|\theta_{obs} - \theta_{mode}| > \frac{\pi}{2}$ **then**
      pattern ← 1;
    end if
  end if
  if *pattern* = 1 **then**
    $\theta \leftarrow \theta_{obs} + \theta_{esc}$;
  else if *pattern* = 2 **then**
    $\theta \leftarrow \theta_{mode} + \theta_{esc}$;
  else if *pattern* = 3 **then**
    $\theta \leftarrow \theta_{mode}$;
  end if

In the NT–Bug2 algorithm, the command direction is regarded as a judgment to estimate the threat level of obstacles to the robot movement. As the condition $\theta_{obs} = \theta_{mode}$ is satisfied, the repulsion force of the obstacle is consistent with the command direction, which means the obstacle would not block the robot's movement along the command line. Therefore, the mobile robot can stop walking around obstacles and move towards the command line directly. In this way, the length of the useless path can be effectively shortened, and the path is smoother and easier to track for the mobile robot.

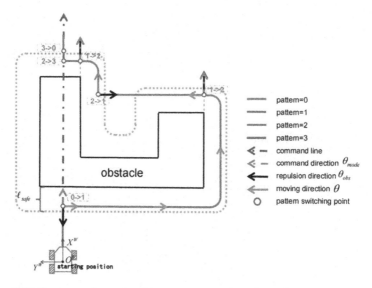

Figure 5.20 The procedure of passing a concave obstacle in the NT-Bug2 algorithm ($\theta_{esc} = \pi/2$).

5.5.3 Motion control

(1) Dynamic error model

For the sake of simplify, we regard the mobile robot as a mass point. We define the global coordinate system as $x^W O^W y^W$ and the robot coordinate system as $x^m O^m y^m$ (Fig. 5.21). We introduce the transformation matrix between the global coordinate system and the robot coordinate system:

$$\mathfrak{R}(\vartheta) = \begin{bmatrix} \cos\vartheta & \sin\vartheta & 0 \\ -\sin\vartheta & \cos\vartheta & 0 \\ 0 & 0 & 1 \end{bmatrix}, \tag{5.79}$$

where ϑ is a rotation angle.

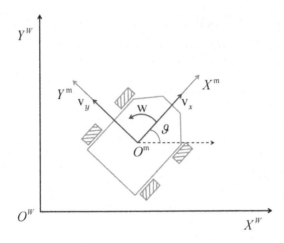

Figure 5.21 Coordinates systems.

The kinematic model of the mobile robot can be expressed as

$$\dot{\mathbf{X}} = \begin{bmatrix} \dot{x} \\ \dot{y} \\ \dot{\vartheta} \end{bmatrix} = \Re^{-1}(\vartheta) \begin{bmatrix} v_x \\ v_y \\ w \end{bmatrix}, \tag{5.80}$$

where $\mathbf{X} = [x, y, \vartheta]^T$ represents the current position of the mobile robot in $x^W O^W y^W$, w is the angular velocity, and v_x and v_y are the velocity components along the x^m-axis and the y^m-axis, respectively. In the same way, we can obtain the kinematic model of the reference position $\mathbf{X}_r = [x_r, y_r, \vartheta_r]^T$. Then, the state error between the current position and the reference position of the mobile robot can be defined as

$$\mathbf{X}_e = \begin{bmatrix} x_e \\ y_e \\ \vartheta_e \end{bmatrix} = \Re(\vartheta) \begin{bmatrix} x_r - x \\ y_r - y \\ \vartheta_r - \vartheta \end{bmatrix}. \tag{5.81}$$

By taking the derivative of Eq. (5.81), we can obtain the dynamic error model:

$$\dot{\mathbf{X}}_e = \begin{bmatrix} \dot{x}_e \\ \dot{y}_e \\ \dot{\vartheta}_e \end{bmatrix} = \begin{bmatrix} -v_x + wy_e + v_{xr}\cos\vartheta_e - v_{yr}\sin\vartheta_e \\ -v_y - wx_e + v_{xr}\sin\vartheta_e + v_{yr}\cos\vartheta_e \\ w_r - w \end{bmatrix}, \tag{5.82}$$

where $[v_{xr}, v_{yr}, w_r]^T$ is the reference velocity. We define

$$\mathbf{u} = \begin{bmatrix} v_{xr} \cos \vartheta_e - v_x \\ v_{yr} \cos \vartheta_e - v_y \\ w_r - w \end{bmatrix}$$

and linearize Eq. (5.82) at the equilibrium point ($\mathbf{X}_e = 0, \mathbf{u} = 0$). Then we can acquire the linearized dynamic error model:

$$\dot{\mathbf{X}}_e = \begin{bmatrix} \dot{x}_e \\ \dot{y}_e \\ \dot{\vartheta}_e \end{bmatrix} = \mathbf{u} + \begin{bmatrix} 0 & w_r & -v_{yr} \\ -w_r & 0 & v_{xr} \\ 0 & 0 & 0 \end{bmatrix} \begin{bmatrix} x_e \\ y_e \\ \vartheta_e \end{bmatrix}. \tag{5.83}$$

(2) Model predictive controller

The task of trajectory tracking is to steer the mobile robot to track the reference trajectory, that is, to find an optimal control input $\mathbf{u}$, so that

$$\lim_{t \to \infty} |\mathbf{X}_r - \mathbf{X}| < \psi, \tag{5.84}$$

where ψ is a small neighborhood containing the origin.

In this system, the reference trajectory consists of two parts: a command trajectory and an obstacle avoidance trajectory. In the former, the moving direction of the mobile robot is given by the operator, while in the latter it is calculated by the NT-Bug algorithm. According to the dead reckoning algorithm, we can deduce the reference trajectory based on the moving direction and the linear velocity. At every sample time, the reference position can be obtained by

$$x_r(k+1) = x_r(k) + \Delta x = x_r(k) + \mathcal{T} v_{xr}$$
$$= x_r(k) + \mathcal{T} v_{linear} \cos \theta, \tag{5.85}$$
$$y_r(k+1) = y_r(k) + \Delta y = y_r(k) + \mathcal{T} v_{yr}$$
$$= y_r(k) + \mathcal{T} v_{linear} \sin \theta, \tag{5.86}$$
$$\vartheta_r(k+1) = \vartheta_r(k) + \Delta \vartheta = \vartheta_r(k) + \mathcal{T} w, \tag{5.87}$$

where $\mathbf{X}_r(0) = [x_r(0), y_r(0), \vartheta_r(0)]^T = \mathbf{X}(0)$, Δx and Δy represent the displacement increment along the x^W- and y^W-axes, respectively, $\Delta \vartheta$ represents the angle increment, and $\mathcal{T}$ is the sampling period. In addition, the angle increment in Eq. (5.87) is zero.

MPC is used for tracking control in the system. By transferring the control problem into a minimization problem of the cost function and calculating it online, a control sequence can be acquired. The first control action of the control sequence is the optimal control input applied to the system.

We rewrite the dynamic error model (5.83) in a discrete-time form:

$$\mathbf{X}_e(k+1) = \mathcal{A}(k)\mathbf{X}_e(k) + \mathcal{B}(k)\mathbf{u}(k), \tag{5.88}$$

where

$$\mathcal{A} = \begin{bmatrix} 1 & \mathcal{T}\mathbf{w}_r & -\mathcal{T}\mathbf{v}_{yr} \\ -\mathcal{T}\mathbf{w}_r & 1 & \mathcal{T}\mathbf{v}_{xr} \\ 0 & 0 & 1 \end{bmatrix},$$

$$\mathcal{B} = \begin{bmatrix} \mathcal{T} & 0 & 0 \\ 0 & \mathcal{T} & 0 \\ 0 & 0 & \mathcal{T} \end{bmatrix}.$$

According to Eq. (5.88), we can define the cost function as

$$\mathbb{F}(\mathbf{X}_e, \mathbf{u}) = \sum_{i=1}^{\Phi} \mathbf{X}_e^T(k+i|k)\Lambda_Q\mathbf{X}_e(k+i|k)$$
$$+ \sum_{i=0}^{\Psi-1} \Delta\mathbf{u}^T(k+i|k)\Lambda_R\Delta\mathbf{u}(k+i|k), \tag{5.89}$$

where Φ and Ψ represent the prediction horizon and control horizon, respectively, satisfying $0 \leq \Psi \leq \Phi$ and $1 \leq \Phi$, Λ_Q and Λ_R are weighting matrices, which are positive and symmetric, $\mathbf{X}_e(k+i|k)$ is the prediction state predicted at sample time k, and $\mathbf{u}(k+i-1|k) + \Delta\mathbf{u}(k+i|k) = \mathbf{u}(k+i|k)$. The optimal control input is obtained by solving the minimization of the cost function with constraints:

$$\hbar = \min_{\mathbf{u}} \quad \mathbb{F}(\mathbf{X}_e, \mathbf{u}), \tag{5.90}$$

$$s.t. \quad \mathbf{X}_e(k) \in [\mathbf{X}^{\min}, \mathbf{X}^{\max}],$$
$$\mathbf{u} \in [\mathbf{u}^{\min}, \mathbf{u}^{\max}],$$
$$\Delta\mathbf{u} \in [\Delta\mathbf{u}^{\min}, \Delta\mathbf{u}^{\max}], \tag{5.91}$$

where $\mathbf{X}^{\min}$ and $\mathbf{X}^{\max}$ are the state constraints, $\mathbf{u}^{\min}$ and $\mathbf{u}^{\max}$ are the input control constraints, and $\Delta\mathbf{u}^{\min}$ and $\Delta\mathbf{u}^{\max}$ are the input increment constraints.

We define the following vectors:

$$\mathfrak{X}_k = [\mathbf{X}_e(k+1|k), \mathbf{X}_e(k+2|k)\ldots, \mathbf{X}_e(k+\Phi|k)]^T, \qquad (5.92)$$

$$\mathfrak{U}_k = [\mathbf{u}(k|k), \mathbf{u}(k+1|k)\ldots, \mathbf{u}(k+\Psi-1|k)]^T, \qquad (5.93)$$

$$\Delta\mathfrak{U}_k = [\Delta\mathbf{u}(k|k), \Delta\mathbf{u}(k+1|k)\ldots, \Delta\mathbf{u}(k+\Psi-1|k)]^T. \qquad (5.94)$$

Then, the predicted output of Eq. (5.88) can be denoted as

$$\mathfrak{X}_k = \Theta\Delta\mathfrak{U}_k + \varpi + \xi, \qquad (5.95)$$

where

$$\Theta = \begin{bmatrix} \mathcal{B}(k|k-1) & \cdots & 0 \\ \vdots & \ddots & \vdots \\ \vdots & \ddots & \vdots \\ \mathcal{B}(k+\Phi-1|k-1) & \cdots & \mathcal{B}(k+\Phi-1|k-1) \end{bmatrix},$$

$$\varpi = \begin{bmatrix} \mathcal{A}(k|k-1)\mathbf{X}_e(k|k-1) \\ \vdots \\ \mathcal{A}(k+\Phi-1|k-1)\mathbf{X}_e(k+\Phi-1|k-1) \end{bmatrix},$$

$$\xi = \begin{bmatrix} \mathcal{B}(k|k-1)\mathbf{u}(k-1) \\ \vdots \\ \mathcal{B}(k+\Phi-1|k-1)\mathbf{u}(k-1) \end{bmatrix}.$$

Based on Eq. (5.95), the optimized objective function (5.90) can be converted to

$$\hbar = \min \, (\Theta\Delta\mathfrak{U}_k + \varpi + \xi)^T \tilde{\Lambda}_Q (\Theta\Delta\mathfrak{U}_k + \varpi + \xi) + \Delta\mathfrak{U}_k^T \tilde{\Lambda}_R \Delta\mathfrak{U}_k, \qquad (5.96)$$

$$s.t. \quad \Theta\Delta\mathfrak{U}_k + \varpi + \xi \in [\mathfrak{X}^{\min}, \mathfrak{X}^{\max}],$$

$$\Delta\mathfrak{U}_k \in [\Delta\mathfrak{U}^{\min}, \Delta\mathfrak{U}^{\max}],$$

$$\mathfrak{U}_{k-1} \in [\mathfrak{U}^{\min}, \mathfrak{U}^{\max}],$$

$$\mathfrak{U}_{k-1} + \tilde{I}\Delta\mathfrak{U}_k \in [\mathfrak{U}^{\min}, \mathfrak{U}^{\max}], \qquad (5.97)$$

where $\tilde{\Lambda}_Q \in \mathcal{R}^{3\Phi \times 3\Phi}$ and $\tilde{\Lambda}_R \in \mathcal{R}^{3\Psi \times 3\Psi}$ are appropriate dimensional matrices, $\mathfrak{X}^{\min}$ and $\mathfrak{X}^{\max}$ are the state constraints, $\mathfrak{U}^{\min}$ and $\mathfrak{U}^{\max}$ are the input control constraints, $\Delta\mathfrak{U}^{\min}$ and $\Delta\mathfrak{U}^{\max}$ are the input increment constraints, and

$$\tilde{I} = \begin{bmatrix} I & 0 & \cdots & 0 \\ I & I & \cdots & 0 \\ \vdots & \vdots & \ddots & \vdots \\ I & I & \cdots & I \end{bmatrix} \in \mathcal{R}^{3\Psi \times 3\Psi}.$$

To solve the optimization function, Eq. (5.96), we transform it into a quadratic programming (QP) problem:

$$\min \frac{1}{2} \Delta\mathfrak{U}_k^T \varphi \Delta\mathfrak{U}_k + \rho^T \Delta\mathfrak{U}_k, \tag{5.98}$$

$$s.t. \quad \Delta\mathfrak{U}^{\min} \leqslant \Delta\mathfrak{U}_k \leqslant \Delta\mathfrak{U}^{\max},$$

$$\begin{bmatrix} -\tilde{I} & \tilde{I} & -\Theta & \Theta \end{bmatrix} \Delta\mathfrak{U}_k \leqslant \begin{bmatrix} -\mathfrak{U}^{\min} + \mathfrak{U}_{k-1} \\ \mathfrak{U}^{\max} - \mathfrak{U}_{k-1} \\ -\mathfrak{X}^{\min} + \varpi + \xi \\ \mathfrak{X}^{\max} - \varpi - \xi \end{bmatrix}, \tag{5.99}$$

where

$$\varphi = \begin{bmatrix} (\Theta^T \tilde{\Lambda}_Q \Theta + \tilde{\Lambda}_R) & \Theta^T \tilde{\Lambda}_Q \\ \tilde{\Lambda}_Q \Theta & \tilde{\Lambda}_Q \end{bmatrix},$$

$$\rho = \begin{bmatrix} 2\Theta^T \tilde{\Lambda}_Q (\varpi + \xi) \\ 2\tilde{\Lambda}_Q (\varpi + \xi) \end{bmatrix}.$$

5.6. APF-based hybrid shared control

APF is an outstanding robotic motion planning algorithm which is usually used to solve the obstacle avoidance problem, proposed by Khatib [32]. Its core is to construct the repulsive force field around the obstacle and the attractive force field around the target position. In this way, the mobile robot is able to avoid collisions while moving towards the target.

In this subsection, based on the APF method in Ref. [10], a hybrid shared control with EMG-based component is developed to avoid obstacles and to improve the bidirectional human–robot perception using force feedback. This force feedback provides the human partner with the awareness to skillfully control the mobile robot when it gets close to the obstacles. Fig. 5.22 shows the framework of the system.

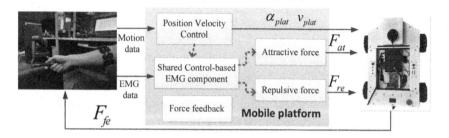

Figure 5.22 Framework of mobile service robots.

As for the leader side, the human partner wears an EMG sensor and moves a haptic device to teleoperate the mobile robot. The EMG sensor is used to capture EMG signals to reflect muscle activation. The haptic device sends positions and velocities to the remote mobile platform by a movable stylus in the Cartesian workspace.

As for the remote robot, it contains a mobile platform with four omnidirectional wheels and is controlled in teleoperated mode. A hybrid shared control scheme with force feedback is proposed for the mobile platform to achieve obstacle avoidance and enable the human partner to adapt their control intention.

5.6.1 Motion control

(1) Dynamics of the mobile platform

Fig. 5.23 shows the configuration of the mobile platform. It can be seen that the mobile platform contains a body and four omnidirectional wheels. For the omnidirectional wheel [33], its velocity along the x-axis $v_{xs,i}$ can be defined as

$$v_{xs,i} = v_{w_i} + v_i \frac{1}{\sqrt{2}}, \tag{5.100}$$

where v_{w_i} represents the velocity of the ith omnidirectional wheel and v_i denotes the velocity of roller i, $i = 1, 2, 3, 4$. Considering the difference in relative positions for four wheels, the velocities along the x-axis are represented in different forms. We have

$$\begin{cases} v_{xs,1} = v_{t,x} - wL_a, \\ v_{xs,2} = v_{t,x} + wL_a, \\ v_{xs,3} = v_{t,x} - wL_a, \\ v_{xs,4} = v_{t,x} + wL_a, \end{cases} \tag{5.101}$$

with

$$L_a = r_{mp}cos\theta_{mp}, \tag{5.102}$$

where $v_{t,x}$ denotes the speed along the x-axis for the mobile platform and w is the angular velocity about the yaw axis.

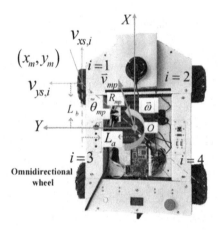

Figure 5.23 The mobile platform.

Correspondingly, the velocities along the y-axis of the mobile platform are

$$\begin{cases} v_{ys,1} = v_i\dfrac{1}{\sqrt{2}} = v_{t,y} + wL_b, \ i = 1, \\ v_{ys,2} = -v_i\dfrac{1}{\sqrt{2}} = v_{t,y} + wL_b, \ i = 2, \\ v_{ys,3} = -v_i\dfrac{1}{\sqrt{2}} = v_{t,y} - wL_b, \ i = 3, \\ v_{ys,4} = v_i\dfrac{1}{\sqrt{2}} = v_{t,y} - wL_b, \ i = 4, \end{cases} \tag{5.103}$$

with

$$L_b = R_{mp}sin\theta_{mp}, \tag{5.104}$$

where $v_{t,y}$ denotes the speed along the y-axis for the mobile platform.

Then, we can obtain the velocities $\{v_{w_i}, i = 1, 2, 3, 4\}$ of the mobile platform:

$$\begin{bmatrix} v_{w_1} \\ v_{w_2} \\ v_{w_3} \\ v_{w_4} \end{bmatrix} = K_{mp} \begin{bmatrix} w \\ v_{t,x} \\ v_{t,y} \end{bmatrix}, \tag{5.105}$$

with

$$K_{mp} = \begin{bmatrix} -L_a - L_b & 1 & -1 \\ L_a + L_b & 1 & 1 \\ -L_a - L_b & 1 & 1 \\ L_a + L_b & 1 & -1 \end{bmatrix}, \tag{5.106}$$

where K_{mp} is a 4×3 matrix.

According to the relationship between velocity and angular velocity, the angular velocity of the mobile platform $\{w_{w_i}, i = 1, 2, 3, 4\}$ can be represented as

$$\begin{bmatrix} w_{w_1} \\ w_{w_2} \\ w_{w_3} \\ w_{w_4} \end{bmatrix} = r_{mp}^{-1} K_{mp} R^{-1} \begin{bmatrix} \dot{\theta} \\ \dot{x} \\ \dot{y} \end{bmatrix}, \tag{5.107}$$

with

$$R = \begin{bmatrix} 1 & 0 & 0 \\ 0 & \cos\theta & -\sin\theta \\ 0 & \sin\theta & \cos\theta \end{bmatrix}, \tag{5.108}$$

where r_{mp} is the radius of the omnidirectional wheel, R represents the rotation matrix between the mobile platform coordinate system and the world coordinate system, x and y are representations of the world frame, and θ denotes the orientation of the mobile platform.

The parameters of the mobile platform can be seen in Table 5.2.

(2) Motion control of the mobile platform

The rotation angle of mobile platform α_{mp} is presented as

$$\alpha_{mp} = \tan(\frac{y_m}{x_m}), \tag{5.109}$$

Table 5.2 Parameters of the mobile platform.

x_m, y_m	Positions of the mobile platform.
$v_{xs,i}$, $v_{sy,i}$	Velocities of the omnidirectional wheel.
θ_{mp}	Angle inclined from the geometric center.
R_{mp}	Distance between the center of mass and the center of the omnidirectional wheel.
$\vec{v}_{mp}$	Velocity of the mobile platform.
$\vec{w}$	Angular velocity of yaw axis rotation.

where y_m and x_m denote the positions along the y-axis and the x-axis of the haptic device, respectively.

Velocity of the mobile platform can be presented as

$$v_{mp} = K_{plat}(z_m - z_{min}) + v_{min}, \qquad (5.110)$$

with

$$K_{plat} = \frac{v_{max} - v_{min}}{z_{max} - z_{min}}, \qquad (5.111)$$

where K_{plat} is a factor to map the velocity of the mobile platform, z_{max} and z_{min} represent the maximum and minimum position of the haptic device along the z-axis, and v_{max} and v_{min} are the maximum and minimum speed of the mobile platform, which can be obtained by a pilot experiment.

The z-axis of the haptic device is used to control the velocity of the mobile platform, and the x- and y-axes are used to describe the motion profile of the mobile platform; therefore, there is a transformation matrix to describe the relationship between the frame of the haptic device and the frame of the mobile platform. The transformation matrix can be represented as follows:

$$R' = \begin{bmatrix} 1 & 0 & 0 \\ 0 & 1 & 0 \\ 0 & 0 & 0 \end{bmatrix}. \qquad (5.112)$$

5.6.2 Hybrid shared control

In this experiment, a hybrid shared control approach based on EMG and APF is exploited to avoid obstacles according to the repulsive force and attractive force and to enhance the human perception of the remote environment based on force feedback of the mobile platform (Fig. 5.24). This shared control method enables the human partner to telecontrol the mobile robot's motion and achieve obstacle avoidance synchronously. Compared with conventional shared control methods, this proposed one provides force feedback based on muscle activation and drives the human partners to update their control intention with predictability.

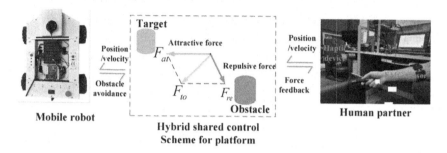

Figure 5.24 Hybrid shared control scheme.

When this hybrid shared control is used and the mobile platform encounters an obstacle, a resultant force (including a repulsive force and an attractive force) will be sent to the mobile platform and drive it away from the obstacle. In this process, the haptic device can receive force feedback (Eq. (5.122)) and provide a stimulus to the human partner. At the same time, the force feedback can make the mobile platform move away from the obstacle.

(1) Processing of EMG signals

In this experiment, we utilize an EMG sensor to capture muscle activation. The EMG signal u_{emg} can be presented as

$$u_{emg} = \sum_{i}^{N} u(i), \quad i = 1, 2, 3, ..., N, \tag{5.113}$$

where $u(i)$ denotes the captured raw EMG signals and N represents the number of channels of the EMG sensor.

In order to obtain the muscle activation accurately, the EMG signals should be filtered through moving average, low pass filter, and envelope.

After filtering of EMG signals, the muscle activation based on EMG signals can be presented as follows:

$$a(i) = \sqrt{\frac{1}{w_{win}} \sum_{i=1}^{w_{win}} u_i^2}, \quad i = 1, 2, ..., w_{win}, \tag{5.114}$$

where $a(i)$ denotes the muscle activation and w_{win} represents the moving window's length.

(2) Combining APF and an EMG-based component in shared control

As for the EMG-based component, inspired by [34], a linear function was developed to describe it. It can be defined as

$$K_{emg} = K^0(a_i - a_{max}) + K^0_{min}, \qquad (5.115)$$

with

$$K^0 = \frac{K^0_{max} - K^0_{min}}{a_{min} - a_{max}}, \qquad (5.116)$$

where K^0 represents the scale parameter of the human factor to adjust the muscle activation, $K^0_{max} \geq K_{emg} \geq K^0_{min}$ is a proportionality coefficient to represent the influence of the EMG-based component, and $a_{max} \geq a_i \geq a_{min}$ denotes the muscle activation [35].

For Eq. (5.115) and Eq. (5.116), when the human partner receives the force feedback through the haptic device, he/she will change his/her manipulation to avoid the obstacle, and the muscle activation (EMG) will change. The EMG can change the values of the repulsive force and the attractive force in the hybrid shared control. In specific, when the mobile platform moves towards the obstacle, the muscle activation transfers to a proportionality coefficient to increase the resultant force to achieve quick avoidance of the obstacle.

As for the APF method, the resultant force sent to the mobile platform contains a repulsive component and an attractive component. The repulsive force propels the platform away from the obstacle, and the attractive force makes the platform move to the target position. The APF of hybrid shared control Q_{to} can be represented as [36,37]

$$Q_{to} = Q_{at} + Q_{re}, \qquad (5.117)$$

with

$$Q_{at} = \frac{1}{2}(\mu_1 + K_{emg})f^2(p, p_{go}), \qquad (5.118)$$

where Q_{at} denotes the hybrid gravitational potential field function, Q_{re} is the hybrid repulsive potential field function, μ_1 is the gravitational gain parameter, and $f(p, p_{go})$ represents the distance from the goal to the mobile platform, where p_{go} is the goal's position. We have

$$
Q_{re} = \begin{cases} \dfrac{1}{2}(\mu_2 + K_{emg})(\dfrac{1}{f(p,p_{ob})} - \dfrac{1}{f_0})^2, & f(p,p_{ob}) \leq f_0, \\ 0, & f(p,p_{ob}) > f_0, \end{cases} \tag{5.119}
$$

where μ_2 is the repulsion gain parameter and f_0 is the influence radius for each obstacle.

Correspondingly, the attractive force can be defined as

$$
\begin{aligned} F_{at} &= -\nabla \, Q_{at} \\ &= (\mu_1 + K_{emg})f(p,p_{go})\frac{\partial f}{\partial p}. \end{aligned} \tag{5.120}
$$

The repulsive force can be defined as

$$
\begin{aligned} F_{re} &= -\nabla \, Q_{re} \\ &= \begin{cases} (\mu_2 + K_{emg})(\dfrac{1}{f(p,p_{ob})} - \dfrac{1}{f_0}) \\ \qquad \dfrac{1}{f^2(p,p_{ob})}\dfrac{\partial f}{\partial p}, & f(p,p_{ob}) \leq f_0, \\ \qquad\qquad 0, & f(p,p_{ob}) > f_0. \end{cases} \end{aligned} \tag{5.121}
$$

(3) Force feedback

There is a distance between the mobile platform and the obstacle in the process of the mobile platform moving to the target position. As shown in Fig. 5.25, when this distance d is less than a safe distance d_s, the mobile platform can generate a force feedback to the human operator through the haptic device (as in Eq. (5.122)). The human operator can change his/her commands to control the mobile platform. We have

$$
F_{fe} = \begin{cases} (K_{fe} + K_{emg})(d_{mw} - d), & d \leq d_s, \\ 0, & d > d_s, \end{cases} \tag{5.122}
$$

where K_{fe} is a positive gain parameter for the platform and d_{mw} and d_s are the maximum warning distance and safe distance, respectively. It can be concluded that when the distance is smaller, the force feedback of the mobile platform is greater.

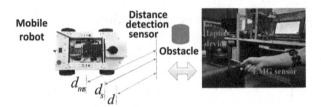

Figure 5.25 Hybrid shared control scheme.

5.7. Experimental case study

5.7.1 Experimental results of the collision avoidance control

(1) Test of neural learning performance

This experiment mainly tests the compensation of the effect caused by the unknown dynamics and the uncertain payload. The Baxter right gripper holds a 1.3-kg payload and the trajectory which controls the right end-effector is (Fig. 5.26)

$$x_d(t) = \begin{bmatrix} 0.6 + 0.1 sin(\frac{2\pi t}{2.5}) \\ -0.4 + 0.3 cos(\frac{2\pi t}{2.5}) \\ 0.2 \end{bmatrix}. \tag{5.123}$$

As for the neural learning network, we will select three nodes for each input dimension, $l = 3^7$ NN nodes for neural networks $\hat{M}(\theta)$ and $\hat{G}(\theta)$, $2l$ NN nodes for neural network $\hat{C}(\theta)$, and $4l$ NN nodes for neural network f. Moreover, all NN weight matrices are initialized as $\mathbf{0}$.

Two comparative experiments are shown, where one is without neural network (Fig. 5.27) and the other one is with neural network (Fig. 5.28). For the experiment without neural learning, we can see the joint angle errors e_θ are relatively high because of the heavy payload. In contrast, for the experiment with neural learning, the joint angle errors e_θ show convergence and become satisfactory after a few cycles.

(2) Test of tracking performance

In this experiment, the end-effector is controlled to move from P_1 : $(0.6, -0.2, 0.2)$ to P_2 : $(0.6, -0.6, 0.2)$ in a line. This experiment can be divided into two small experiments: one without and one with error

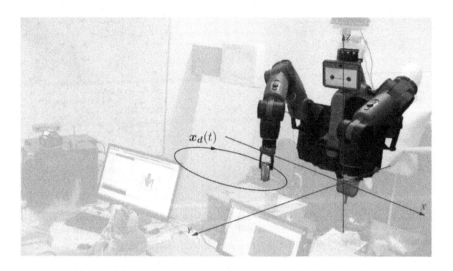

Figure 5.26 Experimental setup for the unknown system dynamics estimation.

transformation method. We choose the parameters in Eq. (5.23) as $\delta = 1$, $\rho_0 = 0.15$, $\rho_\infty = 0.02$, and $p = 1.5$.

The results of these two experiments are shown in Fig. 5.29. We can conclude that the overshoot in the experiment without the error transformation method is much larger and the settling time is much longer than in the experiment with the error transformation method.

(3) Test of collision avoidance

In this experiment, an operator teleoperated a manipulator to move an object from A to B, and there is a potential obstacle in the environment of the Baxter robot, which is shown in Fig. 5.30.

In the sub-experiment without the proposed collision avoidance method, the manipulator collided with the obstacle (Fig. 5.31(a)). However, in the sub-experiment with the proposed collision avoidance method, the manipulator adjusted its position and completed the pick-and-place task (Fig. 5.31(b)).

(4) Test of restoration function

In this experiment, the researcher holds a green joystick which gradually moves closer to the manipulator and then gradually moves away from the manipulator. The results of this experiment are shown in Fig. 5.32. When the joystick moved close to the manipulator, its elbow moved down in

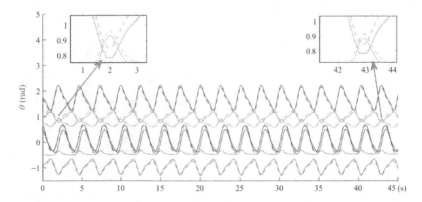

(a) Reference joint angles θ_d and the actual joint angles θ without neural learning. The dashed and solid lines denote reference and actual joint angles, respectively. The lines in different colors denote different joints.

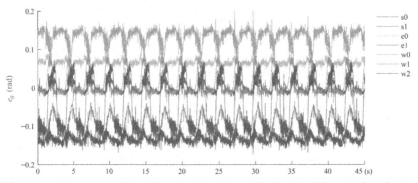

(b) Joint angle error e_θ without the neural learning. The lines in different colors denote different joints.

Figure 5.27 The experiment without neural learning.

order to avoid the joystick. When the joystick moved away, the manipulator automatically restored back to its previous position.

5.7.2 Experimental results of the EEG-based shared control

In this experiment, a human subject was asked to wear an EEG cap connected to the Neuroscan device and sit in front of a monitor which displayed the fusion video. First, the researcher told a random target to the subject, then the subject stared at this target on the screen, and the robot's manipulator moved to pick up this target. In addition, we determined whether the manipulator avoided the obstacles in its trajectories.

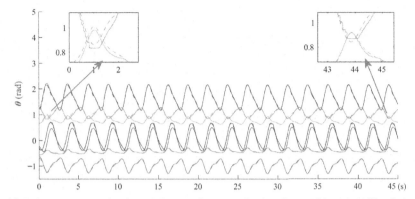

(a) Reference joint angles θ_d and the actual joint angles θ with neural learning. The dashed and solid lines denote reference and actual joint angles, respectively. The lines in different colors denote different joints.

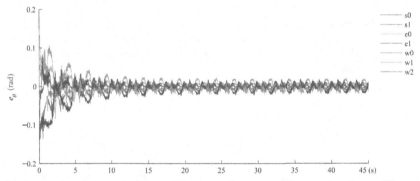

(b) Joint angle error e_θ with the neural learning. The lines in different colors denote different joints.

Figure 5.28 The experiment with neural learning.

Two subjects were invited. Their information is shown in Table 5.3. Subject 1, with SSVEP experience, succeeded in accomplishing all tasks. Subject 2, without any SSVEP experience, failed to accomplish the first task, but succeeded in the next four tasks.

Table 5.3 Subject information and test results.

Subject	Health	Gender	Age	Previous experience	Results
Subject 1	Healthy	Male	24	Yes	5/5
Subject 2	Healthy	Male	23	No	4/5

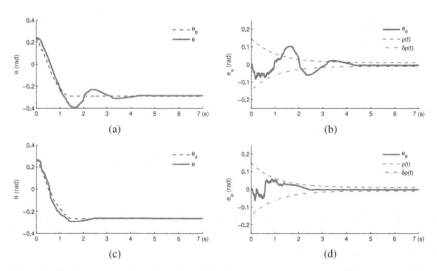

Figure 5.29 Experimental results of the tracking performance test. (a) Joint angle without the error transformation method. (b) Error without error transformation method. (c) Joint angle with the error transformation method. (d) Error with the error transformation method.

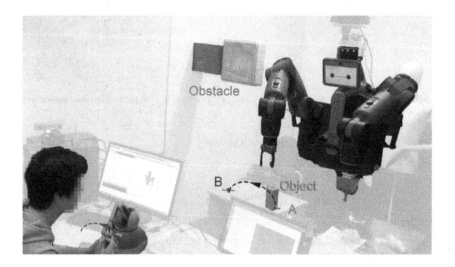

Figure 5.30 Setup of the collision avoidance experiment.

Therefore, we can conclude that the SSVEP paradigm is easy to handle and requires little training. The system can help the disabled to control manipulators easily without extensive training.

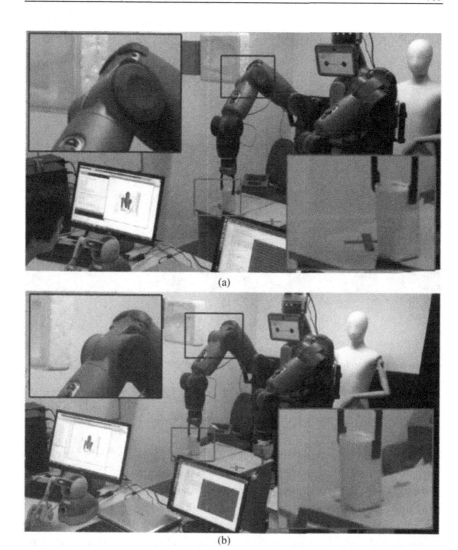

(a)

(b)

Figure 5.31 Video frames of the collision avoidance experiment. (a) Without collision method. (b) With collision method.

5.7.3 Experimental results of the MR-based user interactive path planning

(1) Path planning without virtual obstacles

This experiment tested the actual effect of the proposed path planning algorithm on the known map without virtual obstacles added. In the 16 × 28 m map, one starting point and two target points were set. Moreover,

Figure 5.32 Video frames of the restoring control experiment. (a) $t = 36$ s. (b) $t = 38$ s. (c) $t = 39$ s. (d) $t = 43$ s.

we adopted a dynamic windows approach (DWA) and the improved VFH* method in this map.

The trajectory generated in RViz is shown in Fig. 5.33. We can see that both methods can accomplish the task. Now we consider the changes in velocity and angular velocity of the two algorithms in the process of motion shown in Fig. 5.34(b and c). When adopting the algorithm proposed in this chapter, the maximum and minimum velocity and angular velocity were limited, and the deceleration when obstacles were detected was optimized, so that it would not be reduced to zero directly. In this way, the mechanical motion structure of the mobile robot can be better protected and the moving process can be smoother.

Additionally, as can be seen from Table 5.4, under the premise of setting the same starting point and target points, the improved method in this chapter produces shorter path lengths and shorter robot movement time than the traditional DWA method, and the sum of the distances to the obstacles are higher when the robot passes the same number of obstacles.

The improved algorithm proposed in this chapter was effective in this experiment. The mobile robot could smoothly avoid obstacles and safely

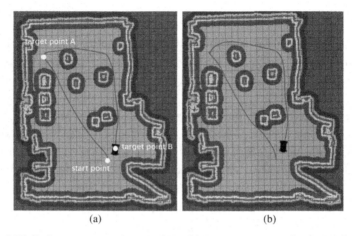

Figure 5.33 Trajectory comparison in RViz. (a) Trajectory using the DWA algorithm. (b) Trajectory using the improved VFH* algorithm.

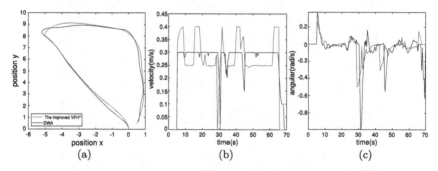

Figure 5.34 The results of different path planning methods were compared in experiment I. (a) Trajectory comparison. (b) Velocity comparison. (c) Angular velocity comparison.

Table 5.4 Experimental results of parameters in the path planning without virtual obstacles.

Parameter	DWA	Improved VFH*
Path length (m)	17.91	17.36
Runtime (s)	64.6	62.2
Passed obstacles	5	5
Sum of the distance from the obstacles (m)	2.95	3.40

reach the specified target point. Besides, the changes in velocity and angular velocity were flexible and the system was robust.

(2) Path planning with virtual obstacles

This part of the experiment showed the path-planning effect of the mobile robot after adding virtual obstacles to the map with HoloLens. We used the same map as in the previous experiment and set a starting point and a target point.

The experimental operation process was as follows: When the mobile robot moved to a certain position, two virtual obstacles were added to the map through gesture control of HoloLens. Then, the path planning algorithm took these two virtual obstacles into account and replanned the path.

The scenes in HoloLens before and after the operation of adding obstacles are shown in Fig. 5.35(a and b).

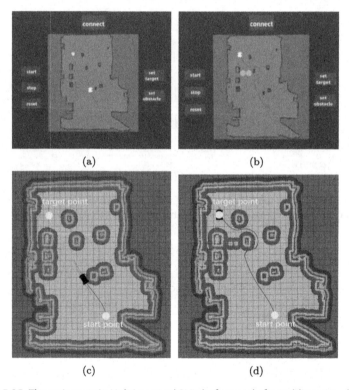

(a)   (b)

(c)   (d)

Figure 5.35 The trajectory in HoloLens and RViz before and after adding virtual obstacles. (a) The scene in HoloLens before adding virtual obstacles. (b) The scene in HoloLens after adding virtual obstacles. (c) The scene in RViz before adding virtual obstacles. (d) The scene in RViz after adding virtual obstacles.

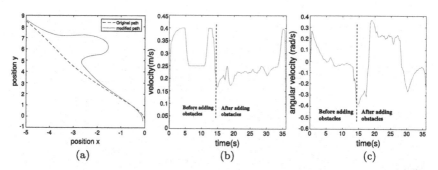

Figure 5.36 The results before and after adding virtual obstacles in experiment II. (a) Trajectory comparison before and after adding obstacles. (b) Change in velocity before and after adding obstacles. (c) Change in angular velocity before and after adding obstacles.

The actual path to avoid virtual obstacles in RViz is shown in Fig. 5.35(c and d). The trajectory comparison of before and after the addition of virtual obstacles is shown in Fig. 5.36(a). It can be seen that in the process of motion, when the mobile robot detected the suddenly added obstacle, it immediately changed the previously planned trajectory, and it then replanned the trajectory according to the new map and moved according to the new trajectory. The changes in velocity and angular velocity are shown in Fig. 5.36(b and c).

(3) Real experiment validation

To verify the actual effect of the system proposed in this chapter in real situations, we tested the system in a real experimental environment using the mobile robot mentioned in this chapter. The experimental environment was limited to a size of 3.5 × 3 m, and three fixed obstacles were placed. This part of the experiment tested a variety of scenes with or without virtual obstacles. The movement process of the mobile robot in the experimental environment is shown in Fig. 5.37.

Consistent with the simulation results, the system can directly control the mobile robot for path planning and set up virtual obstacles by a virtual panel to make the mobile robot avoid the location we do not want to pass and reach the target point successfully. The system can be well applied to situations such as virtual meetings that need to avoid some unreal obstacles.

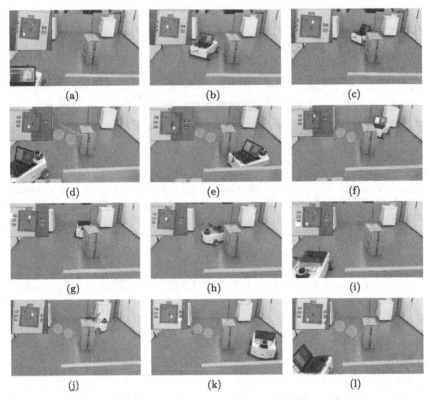

Figure 5.37 Verifying the results of the experiment with the real omnidirectional mobile robot. The top left corner of each picture is the image in HoloLens. Since HoloLens superimposes virtual objects on a real scene, the cabinet in the real scene can be seen in the image. (a–f) The processes of reaching the target point with and without virtual obstacles. The green circles are the virtual obstacles that do not exist in the real scene and only act on the cost map of the mobile robot. (g–l) Processes that return to the starting point with and without virtual obstacles.

5.7.4 Experimental results of no-target obstacle avoidance with sEMG-based control

(1) Experimental setup

In this set of experiments, we apply the proposed NT–Bug algorithm to the shared control system to assist the operator in controlling the motion of the mobile robot. The experimental site is about 3 m × 5 m, as shown in Fig. 5.38(a). The four-wheel OMR is driven by four motors of 90 power and is controlled by a computer with an i7-5500U CPU. A laser radar with a scanning distance of 8 m and an angle accuracy of 0.18 degrees is mounted at the front of the mobile robot platform (Fig. 5.38(b)). The global coordinate system $x^W O^W y^W$ is set up at the starting position of the mobile robot, while its origin O^W is at the center of the mobile robot and the x^W-axis is consistent with the forward direction of the mobile robot. The MPC parameters in the experiments are set as $\Phi = 3$, $\Psi = 2$, $\mathfrak{U}^{max} = -\mathfrak{U}^{min} = [1.5, 1.5, 1.5, 1.5, 1.5, 1.5]^T$, $\mathfrak{U}^{max} = -\mathfrak{U}^{min} = \Delta\mathfrak{U}^{max} = -\Delta\mathfrak{U}^{min} = [10, 10, 10, 10, 10, 10]^T$, $\mathfrak{X}^{max} = -\mathfrak{X}^{min} = [10, 10, 10, 10, 10, 10, 10, 10, 10]^T$, $\tilde{\Lambda}_Q = 2000I$, and $\tilde{\Lambda}_R = 40I$, where $I \in \mathcal{R}^{3\Phi \times 3\Phi}$ is an identity matrix. The minimum and maximum magnitude of the muscle stiffness are set as $\varsigma_{min} = 8$ and $\varsigma_{max} = 80$, respectively. The minimum and maximum linear velocities of the mobile robot are $v_{min} = 0.05$ m/s and $v_{max} = 0.15$ m/s, respectively.

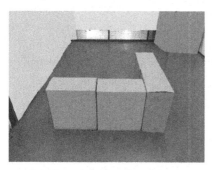 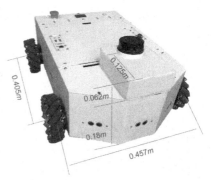

(a) Experiment environment.   (b) The four-wheel omnidirectional mobile robot.

Figure 5.38 Experimental setup.

(2) Comparing NT-Bug1 with NT-Bug2

There is an L-shaped obstacle in the workspace and the operator was asked to control the mobile robot moving forward. It should be noted that in order to demonstrate the advantage of NT-Bug2, the shared control system for collision prevention applied the NT-Bug1 and NT-Bug2 algorithms. So this experiment can be divided into two small experiments. In addition, the parameters of NT-Bug1 are set as $\ell_{safe} = 0.55$, $\theta_{esc} = \pi/2$, and $d_\sigma = 0.002$, while the parameters of NT-Bug2 are set as $\ell_{safe} = 0.5$, $\theta_{esc} = \pi/2$, and $d_\sigma = 0.002$.

The results of the first and second experiments are shown in Figs. 5.39 and 5.40, respectively. In both experiments, the actual trajectories almost

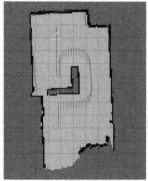

(a) Real-time position of the mobile robot recorded on the grid map.

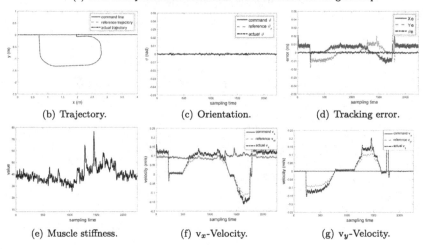

(b) Trajectory.   (c) Orientation.   (d) Tracking error.

(e) Muscle stiffness.   (f) v_x-Velocity.   (g) v_y-Velocity.

Figure 5.39 L-shaped obstacle avoidance using the NT-Bug1 algorithm.

coincide with the reference trajectories, and the tracking errors are small (within 0.01 m), which suggests the trajectory tracking controller works effectively. However, in the first experiment, the path around the contour of the obstacle is longer, and there are more corners on the trajectory, increasing the risk of collision. In contrast, the trajectory in the second experiment is smoother, and the switching frequency between v_x and v_y is lower than that in the first experiment. Comparison of the experimental results shows that the NT–Bug2 algorithm is more efficient for the shared control system. It also can plan a smoother and safer path for the mobile robot.

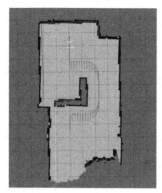

(a) Real-time position of the mobile robot recorded on the grid map.

Figure 5.40 L-shaped obstacle avoidance using the NT-Bug2 algorithm.

(3) Real experiment validation

The last experiment is to validate the effectiveness of the NT-Bug2 algorithm for static and dynamic avoidance. One obstacle is put in the workspace in advance, the other is placed during the process of the experiment. Under the control of the operator, the mobile robot moves left at the first stage and moves forward at the second stage.

The last experimental results are shown in Fig. 5.41. The block of top left corner on the grid map represents a dynamic obstacle, while the block of lower right corner represents a static obstacle. As shown in Fig. 5.41(a), the mobile robot can successfully bypass both static and dynamic obstacles using the NT-Bug2 algorithm. Therefore, the proposed NT-Bug2 algorithm is suitable for both dynamic and static avoidance.

5.7.5 Experimental results of APF-based hybrid shared control
(1) Experimental setup

In this set of experiments, the operator should wear a MYO Armband (THalmic Labs Inc.) and control a Touch X haptic device. The environment is shown in Fig. 5.42. The human partner telecontrols the mobile platform to move and avoid the obstacles in an indoor environment. It is noted that the target position is (350 cm, −40 cm), and operation errors of the human partner in the x- and y-directions are limited to ±5 cm and ±10 cm, respectively. The parameters of shared control are set as $\mu_1 = 100$, $\mu_2 = 100$, $K_{fe} = 1$.

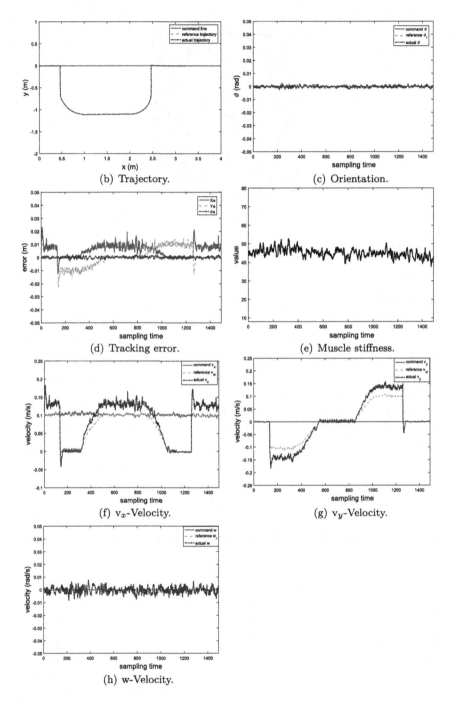

(b) Trajectory.

(c) Orientation.

(d) Tracking error.

(e) Muscle stiffness.

(f) v_x-Velocity.

(g) v_y-Velocity.

(h) w-Velocity.

Figure 5.40 (*continued*)

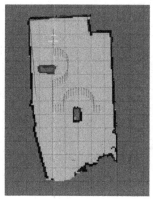

(a) Real-time position of the mobile robot recorded on the grid map.

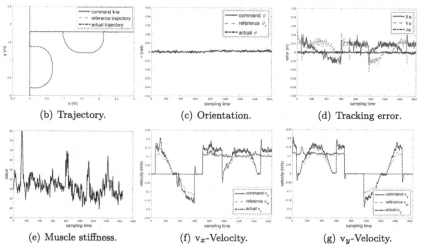

(b) Trajectory.   (c) Orientation.   (d) Tracking error.

(e) Muscle stiffness.   (f) v_x-Velocity.   (g) v_y-Velocity.

Figure 5.41 Static and dynamic obstacle avoidance using the NT-Bug2 algorithm.

In addition, these experiments are conducted in two different conditions: hybrid shared control with an EMG-based component and without EMG-based component. The results without EMG-based component and with EMG-based component are indicated as $-c1$ and $-c2$, respectively.

(2) Obstacle avoidance experiment in a one-obstacle environment

The performance in the one-obstacle environment is shown in Fig. 5.43. Fig. 5.43(a) indicates that when the mobile platform is controlled without EMG-based component, the mobile platform suffers from a small resul-

(a) One-obstacle environment.   (b) Multiobstacle environment.

Figure 5.42 Experimental environment.

tant force in the process of obstacle avoidance. In comparison, the mobile platform with EMG-based component achieves a better performance in obstacle avoidance. Especially, when the mobile platform passes by the obstacle, the haptic device with EMG-based component can receive a larger feedback force than that without EMG-based component (Fig. 5.43(b)). The resultant force and feedback force can drive the mobile platform to move away from the obstacle. Fig. 5.43(c) shows that the method with EMG-based component can achieve obstacle avoidance earlier than that without EMG-based component. As shown in Fig. 5.43(d), the velocity is more continuous in the case with EMG-based component (solid curve) than in the case of without EMG-based component (dashed curve). Fig. 5.43(e) shows that the muscle activation changes abruptly at about 24 s, when the mobile platform gets close to the obstacle. In this sense, we can see that the muscle activation varies with the process of obstacle avoidance and the EMG-based component can enhance the obstacle avoidance performance.

Moreover, from Table 5.5 we can see that the minimum safe distance in the case with EMG-based component is greater than that in the case without EMG-based component. It can be concluded that the proposed method with EMG-based component can achieve a greater minimum safe distance.

(3) Obstacle avoidance experiment in a multiobstacle environment

In this experiment, we aimed to validate the robustness and performance of the proposed method, so a multiobstacle environment was set up.

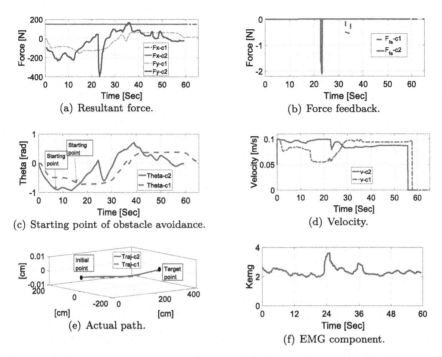

(a) Resultant force.

(b) Force feedback.

(c) Starting point of obstacle avoidance.

(d) Velocity.

(e) Actual path.

(f) EMG component.

Figure 5.43 Performance comparison with/without EMG component in the one-obstacle experiment.

Table 5.5 Total time, displacement of total path traveled, and average minimum safe distance in the one-obstacle experiment.

Parameter	Without EMG	With EMG
Total time (s)	64.9429	58.4510
Total displacement (cm)	179.4354	157.9623
Minimum safe distance (cm)	79.23	80.55

Fig. 5.44 and Table 5.6 show the performance of the proposed method in the multiobstacle environment. It can be seen that the EMG-based method can achieve a better performance in obstacle avoidance than that without EMG-based component, in terms of minimal safe distance, resultant force, and force feedback. Furthermore, the hybrid shared control method can predict the obstacle through the resultant force and feedback force and provides a longer process to compel the mobile platform to move away from the obstacles.

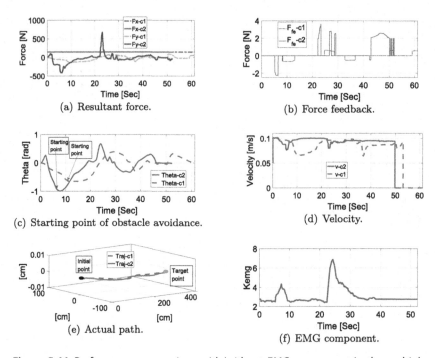

Figure 5.44 Performance comparison with/without EMG component in the multiob-stacle experiment.

Table 5.6 Total time, displacement of total path traveled, and av-erage minimum safe distance in the multiobstacle experiment.

Parameter	Without EMG	With EMG
Total time (s)	60.8470	51.7560
Total displacement (cm)	182.3736	154.8775
Minimum safe distance (cm)	55.08	57.53

Similarly, from Table 5.6, it can be seen that the total time and total displacement are shorter in the case with EMG-based component in comparison with that without EMG-based component.

5.8. Conclusion

In this chapter, first of all, shared control, which can facilitate the task for the human operator and improve the overall efficiency of the system, is briefly introduced. Then, five applications of shared control, namely collision avoidance control, EEG-based shared control, MR-based user interactive path planning, sEMG-based shared control, and APF-based hybrid shared control, are analyzed in detail. They make the manipulators or the mobile robots more intelligent and more easily controllable. Finally, several experiments using these applications are shown to validate their feasibility and robustness.

References

[1] Ronald Lumia, Jon C. Fiala, Albert J. Wavering, An architecture to support autonomy, teleoperation, and shared control, in: Proceedings of the 1988 IEEE International Conference on Systems, Man, and Cybernetics, vol. 1, IEEE, 1988, pp. 472–476.

[2] Jacob W. Crandall, Michael A. Goodrich, Characterizing efficiency of human robot interaction: A case study of shared-control teleoperation, in: IEEE/RSJ International Conference on Intelligent Robots and Systems, vol. 2, IEEE, 2002, pp. 1290–1295.

[3] Weiyong Si, Ning Wang, Chenguang Yang, A review on manipulation skill acquisition through teleoperation-based learning from demonstration, Cognitive Computation and Systems 3 (1) (2021) 1–16.

[4] Xinyu Wang, Chenguang Yang, Hongbin Ma, Long Cheng, Shared control for teleoperation enhanced by autonomous obstacle avoidance of robot manipulator, in: 2015 IEEE/RSJ International Conference on Intelligent Robots and Systems (IROS), IEEE, 2015, pp. 4575–4580.

[5] Jing Luo, Zhidong Lin, Yanan Li, Chenguang Yang, A teleoperation framework for mobile robots based on shared control, IEEE Robotics and Automation Letters 5 (2) (2019) 377–384.

[6] Mario Selvaggio, P. Robuffo Giordano, Fanny Ficuciello, Bruno Siciliano, Passive task-prioritized shared-control teleoperation with haptic guidance, in: 2019 International Conference on Robotics and Automation (ICRA), IEEE, 2019, pp. 430–436.

[7] Huanran Wang, Xiaoping P. Liu, Adaptive shared control for a novel mobile assistive robot, IEEE/ASME Transactions on Mechatronics 19 (6) (2014) 1725–1736.

[8] Sin-Yi Jiang, Chen-Yang Lin, Ko-Tung Huang, Kai-Tai Song, Shared control design of a walking-assistant robot, IEEE Transactions on Control Systems Technology 25 (6) (2017) 2143–2150.

[9] Paul Nadrag, Lounis Temzi, Hichem Arioui, Philippe Hoppenot, Remote control of an assistive robot using force feedback, in: 2011 15th International Conference on Advanced Robotics (ICAR), IEEE, 2011, pp. 211–216.

[10] Yanbin Xu, Chenguang Yang, Xiaofeng Liu, Zhijun Li, A teleoperated shared control scheme for mobile robot based sEMG, in: 2018 3rd International Conference on Advanced Robotics and Mechatronics (ICARM), IEEE, 2018, pp. 288–293.

[11] Xinyu Wang, Chenguang Yang, Zhaojie Ju, Hongbin Ma, Mengyin Fu, Robot manipulator self-identification for surrounding obstacle detection, Multimedia Tools and Applications 76 (5) (2017) 6495–6520.

[12] Anthony A. Maciejewski, Charles A. Klein, Obstacle avoidance for kinematically redundant manipulators in dynamically varying environments, The International Journal of Robotics Research 4 (3) (1985) 109–117.

[13] Jooyoung Park, Irwin W. Sandberg, Universal approximation using radial-basis-function networks, Neural Computation 3 (2) (1991) 246–257.

[14] Chenguang Yang, Zhijun Li, Rongxin Cui, Bugong Xu, Neural network-based motion control of an underactuated wheeled inverted pendulum model, IEEE Transactions on Neural Networks and Learning Systems 25 (11) (2014) 2004–2016.

[15] F.L. Lewis, Suresh Jagannathan, Aydin Yesildirak, Neural Network Control of Robot Manipulators and Non-Linear Systems, CRC Press, 2020.

[16] Long Cheng, Zeng-Guang Hou, Min Tan, Wen-Jun Zhang, Tracking control of a closed-chain five-bar robot with two degrees of freedom by integration of an approximation-based approach and mechanical design, IEEE Transactions on Systems, Man, and Cybernetics 42 (5) (2012) 1470–1479.

[17] Charalampos P. Bechlioulis, George A. Rovithakis, Prescribed performance adaptive control for multi-input multi-output affine in the control nonlinear systems, IEEE Transactions on Automatic Control 55 (5) (2010) 1220–1226.

[18] David R. Hardoon, Sandor Szedmak, John Shawe-Taylor, Canonical correlation analysis: An overview with application to learning methods, Neural computation 16 (12) (2004) 2639–2664.

[19] Chenguang Yang, Hongbin Ma, Mengyin Fu, Advanced Technologies in Modern Robotic Applications, Springer, 2016.

[20] Liqin Zhu, Huitan Mao, Xiang Luo, Jing Xiao, Determining null-space motion to satisfy both task constraints and obstacle avoidance, in: 2016 IEEE International Symposium on Assembly and Manufacturing (ISAM), IEEE, 2016, pp. 112–119.

[21] Parth Rajesh Desai, Pooja Nikhil Desai, Komal Deepak Ajmera, Khushbu Mehta, A review paper on oculus rift – a virtual reality headset, arXiv preprint, arXiv:1408.1173, 2014.

[22] Hideyuki Tamura, Hiroyuki Yamamoto, Akihiro Katayama, Mixed reality: Future dreams seen at the border between real and virtual worlds, IEEE Computer Graphics and Applications 21 (6) (2001) 64–70.

[23] Oren M. Tepper, Hayeem L. Rudy, Aaron Lefkowitz, Katie A. Weimer, Shelby M. Marks, Carrie S. Stern, Evan S. Garfein, Mixed reality with HoloLens: Where virtual reality meets augmented reality in the operating room, Plastic and Reconstructive Surgery 140 (5) (2017) 1066–1070.

[24] Ulrich Iwan, Johann Borenstein, VFH+: Reliable obstacle avoidance for fast mobile robots, in: Proceedings of the 1998 IEEE International Conference on Robotics and Automation, vol. 2, IEEE, 1998, pp. 1572–1577.

[25] Ulrich Iwan, Johann Borenstein, VFH*: Local obstacle avoidance with look-ahead verification, in: Proceedings 2000 ICRA. Millennium Conference. IEEE International Conference on Robotics and Automation. Symposia Proceedings (Cat. No. 00CH37065), vol. 3, IEEE, 2000, pp. 2505–2511.

[26] Todd R. Farrell, et al., A comparison of the effects of electrode implantation and targeting on pattern classification accuracy for prosthesis control, IEEE Transactions on Biomedical Engineering 55 (9) (2008) 2198–2211.

[27] Miguel Simao, Nuno Mendes, Olivier Gibaru, Pedro Neto, A review on electromyography decoding and pattern recognition for human-machine interaction, IEEE Access 7 (2019) 39564–39582.

[28] Di Ao, Rong Song, JinWu Gao, Movement performance of human–robot cooperation control based on EMG-driven hill-type and proportional models for an ankle power-assist exoskeleton robot, IEEE Transactions on Neural Systems and Rehabilitation Engineering 25 (8) (2016) 1125–1134.

[29] David V. Lu, Dave Hershberger, William D. Smart, Layered costmaps for context-sensitive navigation, in: 2014 IEEE/RSJ International Conference on Intelligent Robots and Systems, IEEE, 2014, pp. 709–715.

[30] David Lloyd, Thomas S. Buchanan, A model of load sharing between muscles and soft tissues at the human knee during static tasks, 1996.

[31] Vladimir Lumelsky, Alexander Stepanov, Dynamic path planning for a mobile automaton with limited information on the environment, IEEE transactions on Automatic control 31 (11) (1986) 1058–1063.

[32] Oussama Khatib, A unified approach for motion and force control of robot manipulators: The operational space formulation, IEEE Journal on Robotics and Automation 3 (1) (1987) 43–53.

[33] Alakshendra Veer, Shital S. Chiddarwar, Design of robust adaptive controller for a four wheel omnidirectional mobile robot, in: 2015 International Conference on Advances in Computing, Communications and Informatics (ICACCI), IEEE, 2015, pp. 63–68.

[34] Jing Luo, Chenguang Yang, Ning Wang, Min Wang, Enhanced teleoperation performance using hybrid control and virtual fixture, International Journal of Systems Science 50 (3) (2019) 451–462.

[35] David G. Lloyd, Thor F. Besier, An EMG-driven musculoskeletal model to estimate muscle forces and knee joint moments in vivo, Journal of Biomechanics 36 (6) (2003) 765–776.

[36] Shuzhi Sam Ge, Yan Juan Cui, New potential functions for mobile robot path planning, IEEE Transactions on Robotics and Automation 16 (5) (2000) 615–620.

[37] Johann Borenstein, Yoram Koren, Real-time obstacle avoidance for fast mobile robots, IEEE Transactions on Systems, Man, and Cybernetics 19 (5) (1989) 1179–1187.

CHAPTER SIX

Human–robot interaction in teleoperation systems

6.1. Introduction

At present, robots suffer from problems such as a complex working environment and a low level of intelligence, so robots can only perform simple and repetitive tasks. Due to limited sensing and control mechanisms, the development of fully autonomous robots that can perform tasks in unknown or complex environments is difficult to achieve at present. Therefore, teleoperated robot technology based on human–robot interaction has become one of the main options for solving current problems.

The movement of a teleoperated robot is guided by the operator. In the process of a robot performing a task, the operator acts as a link in the system, through information perception, feedback, and in other ways, not only to know the execution status of the task, but also through giving commands to control the robot. Based on the operator's operating state, operating behavior characteristics and other human factors will affect the working performance of the teleoperated robot. Therefore, for the development of new teleoperated robots, there is a need to improve the perception ability and intelligence of the teleoperated robot and reduce the operator's operating burden and pressure, which has important theoretical and practical significance.

It has been demonstrated that human motion prediction can enhance the quality of the experience and performance in teleoperation [1]. Wang et al. presented a virtual reality method involving a hidden Markov model and K-means clustering to predict a human welder's operation [2]. A Kalman filter method was presented to predict human motion in real-time and applied to a haptic device in teleoperation [3]. These methods used the human motion profile but not the force information in teleoperation. In addition, with the continuous development of sensor technology, human physiological signals are increasingly used in teleoperated tasks. Considerable progress has been made in the research and application of physiological electromyography (EMG) signals. Artemiadis et al. first applied EMG signals as the control signals of an exoskeleton robot [4], which alleviated the need for the design of complex and heavy mechanical mechanisms and sensors in actual production.

In this chapter we introduce a novel prediction method based on auto-regression (AR) to estimate the human motion intention for teleoperation systems. In the proposed method, by updating the AR parameters, human motion prediction can be adapted online. In addition, we use the inter-action forces of human–robot interaction (HRI) to adjust the predicted motion intent. To this end, we propose a virtual force model. The effec-tiveness of the method is verified by experiments.

In addition to the above methods, we will also introduce a linear guid-ance virtual fixture (VF) based on the actual motion characteristics. VF can provide force feedback to the operator and guide the operator to complete the trajectory tracking task more accurately. In order to give full play to the operator's autonomy and decision-making ability in the teleoperation task, the original VF is improved to an adjustable VF based on the EMG signal, where the EMG signal reflects the operator's control intention [5–7].

6.2. AR-based prediction model

In this chapter, we propose a way to adapt to the trajectory of a robot (Fig. 6.1). It can be seen that the robot trajectory can be adjusted according to the prediction of human body movement, which utilizes the interaction profiles between the robot and the human.

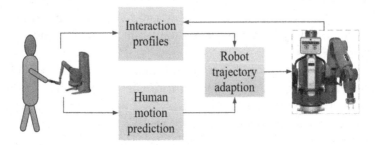

Figure 6.1 Framework of human motion estimation for the teleoperation system.

6.2.1 Problem formulation

In the teleoperation system, we consider the interaction between the haptic device and the human partner in the teleoperation system and the force exerted by the human partner f_h as the input of the system, so we can define the dynamics of the human partner and the haptic device as

$$M_l(x_d)\ddot{x}_d + C_l(x_d, \dot{x}_d)\dot{x}_d + G_l(x_d) = u_l + f_h, \tag{6.1}$$

where M_l and C_l are the inertia matrix and Coriolis and centrifugal matrix of the haptic device, respectively, G_l denotes the gravitational force matrix, x_d is the position in the process of interaction, and u_l denotes the control input of the haptic device.

We implement impedance control for human–robot interaction by designing the control inputs as

$$\begin{aligned}
u_l = {} & C_l(x_d, \dot{x}_d)\dot{x}_d + G_l(x_d) - f_h \\
& + M_l(x_d)\big(\ddot{x}_l - M_{ld}^{-1}\big(D_{ld}(\dot{x}_d - \dot{x}_l) \\
& + K_{ld}(x_d - x_l) - f_h\big)\big),
\end{aligned} \tag{6.2}$$

where M_{ld} denotes the desired inertia matrix of the haptic device, D_{ld} and K_{ld} are the desired damping and stiffness matrices of the haptic device, respectively, and x_l denotes the reference trajectory of the haptic device.

Based on Eqs. (6.1)–(6.2), we have

$$M_{ld}(\ddot{x}_d - \ddot{x}_l) + D_{ld}(\dot{x}_d - \dot{x}_l) + K_{ld}(x_d - x_l) = f_h, \tag{6.3}$$

which is the impedance model of the haptic device. By setting $x_l = 0$ and $K_{ld} = 0$, the haptic device can passively follow the movements of the human partner. But there is still a need for forces from human parts to compensate for the dynamics of the tactile device, so that the tactile device actively cooperates with the human part. This can be observed from the following equation:

$$M_{ld}\ddot{x}_d + D_{ld}\dot{x}_d = f_h. \tag{6.4}$$

On the contrary, if $K_{ld} \neq 0$ and x_l can be designed correctly based on the motion intention of the human partner, the haptic device can move to x_m actively, so it can reduce the effort of the human partner. Therefore, int he next part we will design the reference motion of the haptic device according to the prediction of the motion intention of the human partner.

6.2.2 Prediction model of human motion

The proposed teleoperation system approach can be divided into two parts: the leader part and the follower part. On the leader side, after humans move the haptic device, they can predict the position of the haptic device based on the proposed prediction model. The predicted position is used as the reference position from the robot, and the actual position from the robot will be feedback to the main end to form a closed loop. The specific framework is shown in Fig. 6.2.

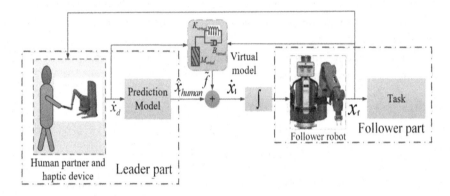

Figure 6.2 Framework of human motion estimation for the teleoperation system.

We suppose that the hand motion of the human partner is a continuous trajectory defined as follows:

$$\dot{x}_{human}(t) = a_0 + a_1 x(t) + a_2 x(t - T) + \dots$$
$$+ a_i x(t - (i-1)T) + \dots + a_p x(t - (p-1)T), \quad (6.5)$$

where T denotes a time step, a_i, $i = 0, 1, 2, \dots, p$, represents the weight, the value of p is determined by the type of human hand motion, and $x(t - iT)$ denotes the information of the ith time step.

For convenience, Eq. (6.5) can be rewritten as follows:

$$\dot{x}_{human}(t) = A^{\mathrm{T}} L(t), \quad (6.6)$$

with

$$A^{\mathrm{T}} = [a_0, \ a_1, \ a_2, \ \dots, \ a_p], \quad (6.7)$$
$$L(t) = [1, \ x(t), \ x(t - T), \dots, \ x(t - (p-1)T)]. \quad (6.8)$$

In order to obtain $\dot{x}_{human}(t)$, we can make the following approximation:

$$\dot{\hat{x}}_{human} = \hat{A}^{\mathrm{T}} L(t) - \alpha_1 \tilde{x}_{human}(t - T), \quad (6.9)$$

with

$$\tilde{x}_{human}(t - T) = \hat{x}_{human}(t - T) - x_{human}(t - T)$$
$$= \hat{x}_{human}(t - T) - x_d(t), \qquad (6.10)$$

where $\hat{A}$ denotes the estimated value of A and α_1 represents a positive scalar.

In order to capture the motion intention of human hand accurately, we propose an update law:

$$\dot{\hat{A}} = -\tilde{x}_{human}(t - T)L(t) - \alpha_2 \tilde{f} L(t), \qquad (6.11)$$

where $\alpha_2 > 0$.

The reference trajectory of the haptic device can be defined as follows:

$$\dot{x}_l = \hat{x}_{human} - \alpha_3 \tilde{f} + \alpha_1 \tilde{x}_{human}(t - T), \qquad (6.12)$$

where α_3 denotes a positive scalar.

6.2.3 Virtual force model

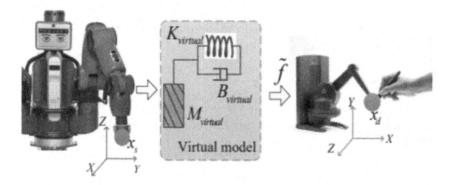

Figure 6.3 An illustration of the virtual model.

The virtual force generated by the motion tracking error between the haptic device and the dependent robot is the interaction force, as shown in Fig. 6.3. We can use virtual forces to represent the admitting model as

$$\tilde{f} = M_{virtual}(\ddot{x}_d - \ddot{x}_f) + B_{virtual}(\dot{x}_d - \dot{x}_f) + K_{virtual}(x_d - x_f), \qquad (6.13)$$

where $M_{virtual}$ is a virtual mass and $B_{virtual}$ and $K_{virtual}$ are the virtual damping and the virtual stiffness, respectively.

During the process of interaction between the teleoperation system and the human partner, virtual forces can provide haptic feedback. In addition, the human hand model can be presented as

$$f_h = K_{human}(x_d - x_{human}),\qquad(6.14)$$

where K_{human} is the control gain of the human partner. It is noted that x_{human} is unknown to the haptic device. In the teleoperation system, the human force is an input of system, so the system of the haptic device can be written as

$$f_h = m(t)\ddot{x}_d + b(t)\dot{x}_d + \tilde{f},\qquad(6.15)$$

where $m(t)$ and $b(t)$ denote the inertia matrix and Coriolis and centrifugal matrix of the haptic device, respectively.

When the motion of the teleoperation system is slow, acceleration $\ddot{x}_d$ and velocity $\dot{x}_d$ of the haptic device can be ignored. Therefore, Eq. (6.14) can be rewritten as

$$f_h = \tilde{f},\qquad(6.16)$$

which indicates that we can obtain the human applied force f_h without a force sensor through the virtual model.

6.2.4 Convergence analysis

In this part, we aim to establish the convergence of the human motion estimation approach. We considering the following Lyapunov function candidate:

$$V = V_1 + V_2 + V_3 + V_4,\qquad(6.17)$$

where

$$V_1 = \frac{1}{2\alpha_2}\,\mathrm{trace}\left(\tilde{A}^{\mathrm{T}}\tilde{A}\right),\qquad(6.18)$$

$$V_2 = \frac{1}{2\alpha_2}\tilde{x}_{hu\,man}^{\mathrm{T}}\tilde{x}_{human}D,\qquad(6.19)$$

where $e = x_d - x_m$, and

$$V_4 = \frac{1}{2}(x_d - x_{hu\,man})^{\mathrm{T}} K_{hu\,man}^{\mathrm{T}}(x_d - x_{human}).\qquad(6.20)$$

For V_1, by taking the derivative with respect to time, we have

$$
\begin{aligned}
\dot{V}_1 &= \frac{1}{\alpha_2} \, \mathrm{trace} \left(\tilde{A}^{\mathrm{T}} \tilde{A} \right) \\
&= \frac{1}{\alpha_2} \, \mathrm{trace} \left(\tilde{A}^{\mathrm{T}} \dot{\tilde{A}} \right),
\end{aligned}
\tag{6.21}
$$

where $\dot{A} = 0$.

Based on Eq. (6.11) and Eq. (6.21), we have

$$
\dot{V}_1 = - \mathrm{trace} \left(\tilde{A}^{\mathrm{T}} \left(\frac{1}{\alpha_2} \tilde{x}_{human}^{\mathrm{T}} (t - T) L(t) + \tilde{f}^{\mathrm{T}} L(t) \right) \right).
\tag{6.22}
$$

Differentiating V_2 with respect to time, we have

$$
\dot{V}_2 = -\frac{1}{\alpha_2} \tilde{x}_{\mathrm{human}}^{\mathrm{T}} \dot{\tilde{x}}_{\mathrm{human}}.
\tag{6.23}
$$

Based on Eq. (6.6) and Eq. (6.9), we have

$$
\dot{\tilde{x}}_{human}(t) = \tilde{A}^{\mathrm{T}} L(t) - \alpha_1 \tilde{x}_{human}(t - T).
\tag{6.24}
$$

Therefore, Eq. (6.23) can be rewritten as

$$
\dot{V}_2 = -\frac{1}{\alpha_2} \tilde{x}_{human}^{\mathrm{T}} \left(\tilde{A}^{\mathrm{T}} L(t) - \alpha_1 \tilde{x}_{human}(t - T) \right).
\tag{6.25}
$$

We assume that the time step T is small and we combine Eq. (6.22) and Eq. (6.25), so we have

$$
\dot{V}_1 + \dot{V}_2 = -\tilde{f}^{\mathrm{T}} \tilde{A}^{\mathrm{T}} L(t) - \frac{1}{\alpha_2} \tilde{x}_{human}^{\mathrm{T}} \alpha_1 \tilde{x}_{hu\,man}.
\tag{6.26}
$$

Differentiating V_3 with respect to time, we have

$$
\dot{V}_3 = \dot{e}^{\mathrm{T}} \left(M_{ld} \ddot{e} + K_{ld} e \right).
\tag{6.27}
$$

Based on the impedance model equation (6.3), we have

$$
\begin{aligned}
\dot{V}_3 &= \dot{e}^{\mathrm{T}} \left(-D_{ld} \dot{e} + \tilde{f} \right) \\
&= -\dot{e}^{\mathrm{T}} D_{ld} \dot{e} + \dot{e}^{\mathrm{T}} \tilde{f}.
\end{aligned}
\tag{6.28}
$$

Differentiating V_4 with respect to time, we have

$$
\dot{V}_4 = \left(x_d - x_f \right)^{\mathrm{T}} K_{human}^{\mathrm{T}} \left(\dot{x}_d - \dot{x}_{human} \right).
\tag{6.29}
$$

Based on Eq. (6.12) and Eq. (6.16), Eq. (6.29) can be rewritten as

$$
\begin{aligned}
\dot{V}_4 &= \tilde{f}^{\mathrm{T}} \left(\dot{x}_d - \dot{x}_{\mathrm{human}} + \dot{\tilde{x}}_{human} \right) \\
&= \tilde{f}^{\mathrm{T}} \left(\dot{x}_d - \dot{x}_l - \alpha_3 \tilde{f} + \alpha_1 \tilde{x}_{hu\,man} + \dot{\tilde{x}}_{human} \right) \\
&= -\alpha_3 \tilde{f}^{\mathrm{T}} \tilde{f} + \tilde{f}^{\mathrm{T}} \dot{e} + \tilde{f}^{\mathrm{T}} \left(\alpha_1 \tilde{x}_{human} + \dot{\tilde{x}}_{human} \right).
\end{aligned}
\tag{6.30}
$$

Based on Eq. (6.24), we have

$$
\dot{V}_4 = -\alpha_3 \tilde{f}^{\mathrm{T}} \tilde{f} + \tilde{f}^{\mathrm{T}} \dot{e} + \tilde{f}^{\mathrm{T}} \tilde{A}^{\mathrm{T}} L(t).
\tag{6.31}
$$

Differentiating V with respect to time, we have

$$
\begin{aligned}
\dot{V} &= \dot{V}_1 + \dot{V}_2 + \dot{V}_3 + \dot{V}_4 \\
&= -\frac{1}{\alpha_2} \tilde{x}_{human}^{\mathrm{T}} \alpha_1 \tilde{x}_{human} - \dot{e}^{\mathrm{T}} D_{ld} \dot{e} - \alpha_3 \tilde{f}^{\mathrm{T}} \tilde{f} \\
&\le 0.
\end{aligned}
\tag{6.32}
$$

Therefore, when $t \to \infty$, $\tilde{x}_{human} \to \infty$, $\dot{e} \to 0$, and $\tilde{f} \to 0$. In this sense, the desired motion of the human hand can be obtained, and the haptic device can track the reference motion x_d.

6.3. EMG-based virtual fixture

This chapter introduces a linear guidance VF based on actual motion characteristics. Adding VF to the teleoperation robot can provide force feedback for the operator, thereby helping the operator to more accurately control the robot to complete the trajectory tracking task. At the same time, the operator's EMG signal is obtained to reflect the operator's intention, and the VF is improved to a regulable virtual fixture based on EMG signals, so that the operator can complete the teleoperation task more autonomously.

6.3.1 Linear flexible guidance virtual fixture

The linear guidance flexible VF (LGFVF) guides the robot as it makes curved movements. Considering the situation where the end-effector of the robot moves freely in Cartesian space, we define P_0 as the end-effector position, and $P_0(x_0(s), y_0(s), z_0(s)) \in R^3$ in Fig. 6.4. The pre-defined tracking curve $P(s)$ can be expressed as

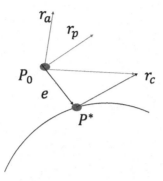

Figure 6.4 The free motion of the robot end-effector in Cartesian space.

$$P_0(s) = (x(s), y(s), z(s)). \tag{6.33}$$

The square distance between P_0 and any point on the tracking curve can be defined as $f(s)$, which can be expressed as

$$f(s) = (x(s) - x_0(s))^2 + (y(s) - y_0(s))^2 + (z(s) - z_0(s))^2. \tag{6.34}$$

Let $f(s)$ take the derivative of s, and set the derivative equal to 0, which can be expressed as

$$df(s)/ds = 0. \tag{6.35}$$

Deriving the equation $f(s)$ yields the minimum point P^* of curve $P(s)$ relative to P_0. Because P^* is the closest point to P_0 on the curve, the tangent direction of $P(s)$ at P^* is the ideal direction of motion for the end-effector, which can be expressed as

$$P'(s(P^*)) = \frac{d}{ds} P(s)|_{s=s(P^*)}, \tag{6.36}$$

$$r_p = \frac{P'(s(P^*))}{\| P'(s(P^*)) \|}, \tag{6.37}$$

where r_p is the ideal motion direction of the robot end-effector.

In LGFVF, the magnitude and direction of the guiding force need to be determined. When the end-effector moves along the tracking curve, r_p can be used as the ideal direction of motion for the robot end-effector, at which point no directing force is required. When the position of the end-effector deviates from the tracking curve, but r_p is selected as the ideal direction of motion and a guiding force is generated in this direction, the problem of the end-effector moving parallel to the tracking curve will occur.

To solve the above problem, the distance (error) between P_0 and r_p is defined as e, which can be expressed as

$$e = P(s(P^*)) - P_0. \tag{6.38}$$

Thus, the new ideal motion direction r_c is defined, which can be expressed as

$$r_c = r_p + k_e e, \tag{6.39}$$

where k_e is the scalar gain factor. When r_c is used as the guiding direction of the virtual fixture, the end-effector will gradually be guided back to the tracking curve, improving the tracking performance of the follower robot.

In this work, a general linear spring model was chosen to describe the guiding forces generated by LGFVF. According to the linear spring model, the feedback force of LGFVF can be described as

$$F_v = k_f e, \tag{6.40}$$

where k_f represents the scalar fixed gain of the virtual guide force and e represents the degree of deviation between the actual trajectory and the desired trajectory. Because the guiding eye of LGFVF is to guide the end effect of the robot to move in the desired direction, e can be replaced with the ideal direction of motion r_c. Then Eq. (6.40) can be written as

$$F_v = k_f r_c, \tag{6.41}$$

where F_v represents the guidance force generated by LGFVF. Considering the inertia factor of the end-effector, when $e > 0$, if the direction of motion of the end-effector is close to the tracking curve, excessive force feedback guidance in this direction can easily cause the robot end to vibrate and reduce the stability margin. Therefore, it is necessary to analyze the relationship between the actual direction of motion r_a and the deviation e. As shown in Fig. 6.5, when the direction of motion of the terminal effector is r_{a4}, the guiding force in that direction will be adjusted as follows:

$$\tilde{r}_c = sign(r_a, e) r_c, \tag{6.42}$$

where

$$sign(r_a, e) = \begin{cases} 1, & cos(r_a, e) \leq 0 \quad \text{or} \quad cos(r_a, r_p) \leq 0, \\ sin(r_a, e), & cos(r_a, e) > 0 \quad \text{and} \quad cos(r_a, r_p) > 0. \end{cases} \tag{6.43}$$

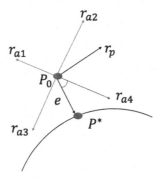

Figure 6.5 Motion direction of the end-effector.

According to Eq. (6.42), when the actual direction of motion is close to the tracking curve in the positive direction (r_{a4}), the magnitude of the guidance force is determined by the angle between r_{a4} and e. When the angle is small, the actual directing force generated is smaller, thus avoiding unstable trajectory vibrations caused by end-effector inertia.

Similarly, Eq. (6.41) can be written as

$$F_v = k_f \tilde{r}_c. \tag{6.44}$$

Thus, for curve tracking in Cartesian free space, the magnitude and direction of the feedback force generated by the linear guided VF can be determined by Eq. (6.44).

6.3.2 Raw sEMG signal pre-processing

EMG signals from the operator's arm are collected through the MYO Armband [8]. Eight sensors in the armband can obtain electrical pulse signals generated by the forearm muscles of the operator, and the operator's arm movements can be analyzed according to changes in the biological signal characteristics of the muscles to determine the operator's intentions [9].

The raw EMG signal needs to be pre-processed to achieve a muscle activity control gain that reflects the operator's intent. The original EMG signal collected from the MYO Armband is shown in Fig. 6.6.

The relationship between the raw EMG signal and muscle excitation can be expressed as

$$e_{raw} = f_a(u), \tag{6.45}$$

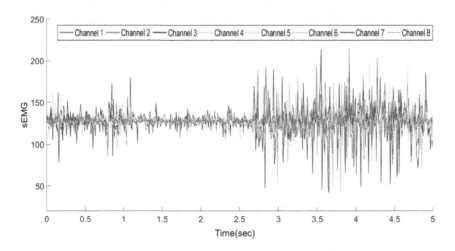

Figure 6.6 Raw EMG signal.

where e_{raw} represents the original EMG signal, u represents muscle excitation, and f_a represents the conversion relationship between e_{raw} and u. As can be seen from Fig. 6.6, the original EMG signal has eight different channels [10]. In order to obtain the EMG signal in the integrated form, the root mean square algorithm is used to integrate the original EMG signal of different channels. This process can be expressed as

$$e(t) = \sqrt{\frac{\sum_{i=1}^{M} e_{\text{rawi}}^2(t)}{M}}, \quad i = 1, 2, 3, \dots M, \qquad (6.46)$$

where e_{rawi} represents the original EMG signal for channel i and $M = 8$ and T represent the total number of channels and sampling time of the MYO Armband, respectively. As shown in Fig. 6.7, the root mean square algorithm is used to integrate the original EMG signals of different channels into a common channel.

Because the EMG signal still has high-frequency noise after integration, we optimize it with an exponentially weighted moving average (EWMA) filtering algorithm to obtain a smoother EMG signal. The process can be expressed as

$$E(t) = \begin{cases} e(t), & t = 1, \\ \beta E(t-1) + (1-\beta)e(t), & t > 1, \end{cases} \qquad (6.47)$$

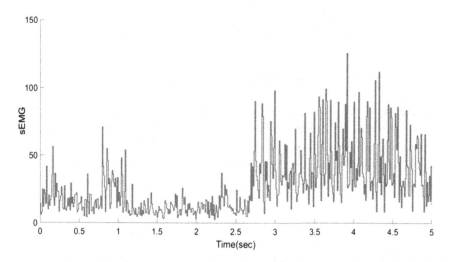

Figure 6.7 Integrated EMG signal.

where β represents the weight of the influence of past measurements on current measurements and $E(t)$ represents the EMG signal filtered by the EWMA filtering algorithm at time t. As shown in Fig. 6.8, after the EWMA algorithm, the EMG signal becomes smoother and more stable [11].

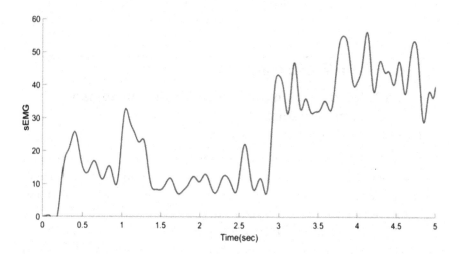

Figure 6.8 Processed EMG signal.

The relationship between muscle activity and processed EMG signals can be expressed as

$$a(t) = \frac{e^{AE(t)} - 1}{e^A - 1},$$
(6.48)

where $A \in (-3, -0)$ is the selected nonlinear shape coefficient and $a(t)$ represents muscle activity. When the control gain is normalized within a reasonable range, it is possible to obtain a control gain that reflects changes in muscle activity and ensures system stability. The linear relationship between the EMG signal and the control gain can be expressed as

$$G(t) = G_{\min} + (G_{\max} - G_{\min}) \frac{a(t) - a_{\min}}{a_{\max} - a_{\min}},$$
(6.49)

where $G(t)$ is the muscle activity gain based on experimental experience, G_{max} and G_{min} are the maximum and minimum gain values, which are used to obtain control gain within a reasonable range and ensure the stability of the control system, and a_{max} and a_{min} are the maximum and minimum values of muscle activity $a(t)$ and are also pre-determined based on experimental experience. After obtaining a control gain that reflects changes in human arm muscle activity, we replace the fixed gain k_f in Eq. (6.44) with $G(t)$, which can be expressed as

$$F_v = G(t)\tilde{r}_c,$$
(6.50)

where $G(t)$ can be independently adjusted by the operator by changing muscle activity to achieve personalized control of LGFVF.

6.4. Experimental case study

6.4.1 Teleoperation experiments based on human motion prediction

(1) Experimental setup

The experimental platform is used to verify the effectiveness of the developed method. The haptic device uses a Touch X joystick, and signals are transmitted from the device to a seven-joint Baxter robot. Humans hold the Touch X device to manipulate the Baxter robot. All devices are controlled using a Robot Operating System (ROS). The Baxter robot is controlled using a PD controller with $K_p = diag[150, 80, 25, 10, 3, 3, 2.5]$ and $K_d = diag[7, 6, 5, 2, 0.8, 0.8, 0.015]$, $\alpha_1 = 0.0001$, $\alpha_2 = 0.00001$, and

$\alpha_3 = 0.2$. The mean absolute error (MAE) is used to quantify the performance of the developed method.

(2) Experiment I

Experiment I included three different movements. Fig. 6.9 shows the tracking performance results with a delay of 20 ms.

Fig. 6.9(a) shows that the follower cannot accurately track the leader throughout the process without making a prediction to perform teleoperation. In contrast, teleoperation enables improves the trajectory tracking performance in predicted scenarios. Fig. 6.9(b) shows the tracking error with and without prediction. It is not difficult to see that the dependent robot achieves a small tracking error by predicting the intention of the human movement.

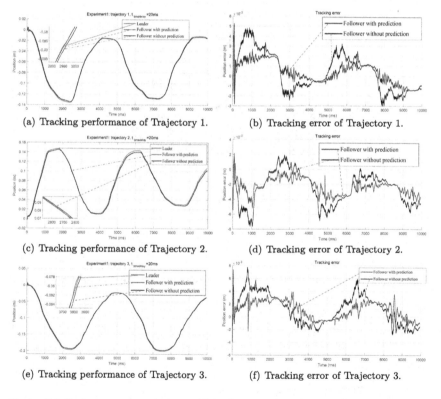

(a) Tracking performance of Trajectory 1.   (b) Tracking error of Trajectory 1.

(c) Tracking performance of Trajectory 2.   (d) Tracking error of Trajectory 2.

(e) Tracking performance of Trajectory 3.   (f) Tracking error of Trajectory 3.

Figure 6.9 Performance comparison with/without prediction in Experiment 1.

To verify the robustness of the proposed method, we also conducted experiments with two other movements. Figs. 6.9(c)–6.9(f) show the tracking results of tracks 2 and 3. Compared to the tracking performance of track 1, the results of tracks 2 and 3 are similar. It can be concluded that the follower can update its trajectory online based on predictions of human movement and accurately track the leader in the presence of a time delay.

Table 6.1 Comparison of MAE values for Trajectories 1, 2, and 3 in Experiment I with a time delay of 20 ms.

Trajectory	Without prediction (m)	With prediction (m)
Trajectory 1	0.0017	8.0279e−4
Trajectory 2	0.0027	0.0020
Trajectory 3	0.0022	0.0012

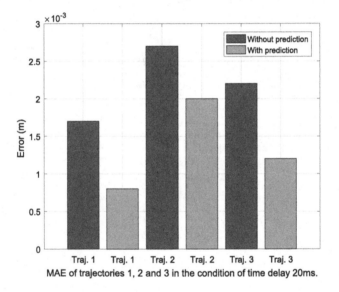

MAE of trajectories 1, 2 and 3 in the condition of time delay 20ms.

Figure 6.10 Comparison of MAE values with/without prediction in Experiment I.

Table 6.1 and Fig. 6.10 show the tracking performance of Trajectories 1, 2, and 3 at a time delay of 20 ms. It can be concluded that the proposed prediction-based approach is capable of achieving a smaller tracking error than the method with no prediction.

(3) Experiment II

In this section, we demonstrate the performance of the proposed method under different time delay conditions. Fig. 6.11(a) illustrates the tracking performance without prediction at a time lag of 20 ms. There is a larger error when the follower tracks the leader with time delays of 50 ms (Fig. 6.11(c)) and 80 ms (Fig. 6.11(e)). The tracking error increases with increasing time delay. In this sense, a teleoperation system would not work without human motion predictions.

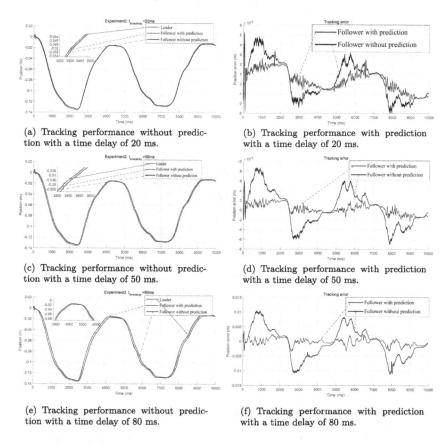

(a) Tracking performance without prediction with a time delay of 20 ms.

(b) Tracking performance with prediction with a time delay of 20 ms.

(c) Tracking performance without prediction with a time delay of 50 ms.

(d) Tracking performance with prediction with a time delay of 50 ms.

(e) Tracking performance without prediction with a time delay of 80 ms.

(f) Tracking performance with prediction with a time delay of 80 ms.

Figure 6.11 Performance comparison with/without prediction in Experiment II.

Using the proposed prediction-based method, the trajectory of the follower station remains close to the trajectory of the leader. Figs. 6.11(b), 6.11(d), and 6.11(f) show that the proposed method can achieve similar performance with time delays of 20 ms, 50 ms, and 80 ms, showing that the performance of the proposed method is independent of the delay.

Table 6.2 and Fig. 6.12 present a performance comparison of the teleoperation system based on MAE values. It can be seen that the MAE values without prediction are larger than those with prediction.

Table 6.2 Comparison of MAE values in Experiment II with time delays of 20 ms, 50 ms, and 80 ms.

Time delay	Without prediction (m)	With prediction (m)
20 ms	0.0017	8.0279e−4
50 ms	0.0031	7.4042e−4
80 ms	0.0045	7.9180e−4

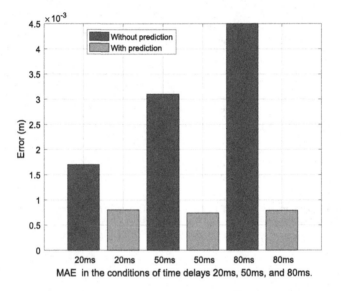

Figure 6.12 Comparison of MAE values with/without prediction in Experiment II.

6.4.2 Experiment of EMG-based virtual fixture

We performed experiments with the Touch X–Baxter teleoperation system. In this teleoperation system, the human operator can control the movement of the follower device (the Baxter robot) through the leader device (the Touch X device), thus achieving precise position control. The experiment consists of two parts. First, we verified the effectiveness of LGFVF. The second part of the experiment verifies the effectiveness of the tunable LGFVF algorithm by adding a muscle electrical signal acquisition and processing module to the teleoperation system and introducing a muscle activity control gain that reflects the decision-making ability of human operators in the LGFVF module.

(1) Validation of LFGVF

In this experiment, we designed a specific trajectory tracking task that requires operator control to move the follower end-effector along a specified trajectory. By comparing the tracking effect of the specified trajectory under the manual control of the operator and LFGVF control, the effectiveness of the LGFVF algorithm is verified. As shown in Fig. 6.13, the pre-defined tracking trajectory is designed as a semicircle with a radius of 2 cm on the xz-plane, starting at $(0,0)$ and ending at $(4,0)$. This task requires the operator to control the movement of the follower actuator from the start point to the endpoint along a pre-defined trajectory.

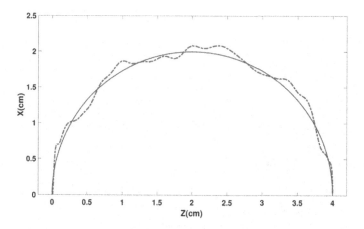

Figure 6.13 End-effector trajectory of the follower in manual control.

The operator first manually controls the movement of the follower robot, unguided by any force feedback in the process. The experimental results are shown in Fig. 6.13, where the solid blue line is a specified trajectory generated by the software in advance, while the dotted red line is the movement trajectory of the follower actuator generated by the operator controlling the movement of the subordinate Baxter robot. When the robot is manually controlled by the operator, there is a large deviation between the follower end actuator trajectory and the pre-defined trajectory. Therefore, in the case of manual control, the accuracy of teleoperation tasks cannot be guaranteed.

Fig. 6.16 shows the magnitude of the force feedback received by the operator using the linear guidance virtual fixture algorithm throughout the task.

It can be seen that the magnitude of the force feedback is positively correlated with the deviation between the trajectory of the follower end-effector and the pre-defined trajectory. The greater the deviation, the stronger the virtual fixture's guiding effect on the operator (Fig. 6.14).

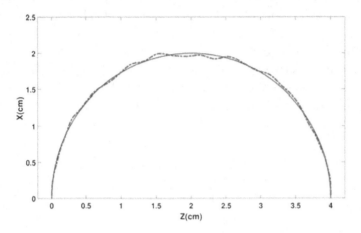

Figure 6.14 Follower end-effector trajectory in virtual fixture control.

(2) Validation of regulable LGFVF

In order to verify the effectiveness of the regulable VF, we carried out teleoperation experiments under three control modes: (1) manual control; (2) original VF control; and (3) regulable VF control.

In this work, the tracking trajectory is set to two interconnected semi-circles with a radius shown in Fig. 6.15, one of which is in the positive direction of the x-axis and the other is in the negative direction of the x-axis. Unlike previous tasks, obstacles and position interference factors have been added to the track tracking task. The obstacle is a solid circle with a radius of 0.4 cm, the center of which is at the top of the first semicircle in the path. The operator needs to control the follower end-effector to bypass obstacles and avoid collisions with them. In addition, we added a sinusoidal interference of $\Delta X = 0.1\sin(3t)$ cm in the second semicircle, which affects the position of the follower end-effector.

Fig. 6.15 shows the trajectory tracking effect under manual operator control. It is not difficult to see that the operator consciously controls the follower end-effector to bypass the obstacle in the first half of the path. However, there is a large tracking deviation throughout the process, especially in the second half of the path, and the trajectory of the follower

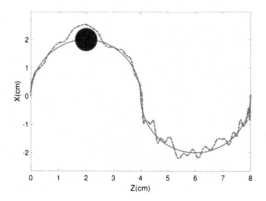

Figure 6.15 Manual control with obstacle and under interference.

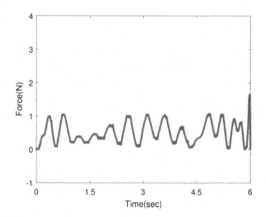

Figure 6.16 Force feedback in VF control.

end-effector fluctuates greatly due to the influence of sinusoidal interference.

Fig. 6.17 shows the trajectory tracking effect under LGFVF guidance. The parameters of the VF are consistent with previous experiments. As can be seen from the figure, under the guidance of LGFVF's force feedback, not only the tracking performance is improved, but the trajectory tracking error of the whole process is significantly reduced. However, due to the force feedback constraints of LGFVF, although the operator consciously controlled the follower end-effector to bypass the obstacle, it still failed to avoid contact with the obstacle. It can be seen that the influence of LGFVF not only improves the accuracy of teleoperation, but also reduces the operator's ability to control autonomously.

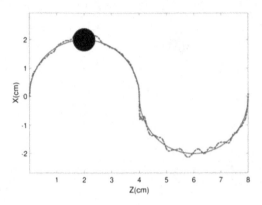

Figure 6.17 Manual control with obstacle and under interference.

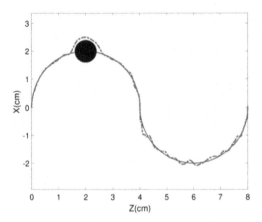

Figure 6.18 Regulable VF control with obstacle and under interference.

Fig. 6.18 shows the trajectory tracking performance under adjustable LGFVF control. The operator can adjust the control gain of LGFVF by changing muscle activity. As can be seen from the figure, the operator controls the gain of the VF within the appropriate range in the first half of the path so that the follower end-effector can more accurately track the pre-defined trajectory. When approaching an obstacle, the operator adjusts the gain of LGFVF by varying muscle activity to reduce the degree of assistance of LGFVF. At this point, the operator controls the end-effector successfully bypassing the obstacle. In the second half of the path (back half of the circle), the operator increases the degree of assistance of LGFVF by changing muscle activity, so that it still has good tracking performance despite sinusoidal interference.

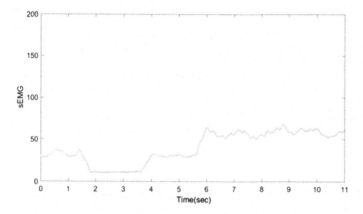

Figure 6.19 Muscle activity of the operator in regulable virtual fixture control mode.

Fig. 6.19 shows the changes in muscle activity of the operator in the regulable virtual fixture control mode. In comparison to Figs. 6.15–6.18, it can be seen that the proposed teleoperation control scheme based on the regulable virtual fixture algorithm performs best in terms of operational performance and operational safety in teleoperation tasks.

6.5. Conclusion

In this chapter, we have introduced a prediction method for estimating human motion based on an AR model. In this method, human body movement can be predicted by updating the parameters of the AR model, and then the robot's trajectory can be adjusted automatically. Furthermore, techniques such as haptic feedback are used. We also introduce a linear guidance VF based on sEMG signals. The aim is to guide the robot through a trajectory more accurately by providing force feedback to the operator. Experiments including several different tasks were carried out using a Baxter robot. Good performance with a small position error and a low contact force was obtained, indicating this method may find applications in human–robot systems such as teleoperation systems.

References
[1] T. Aykut, C. Zou, J. Xu, D.V. Opdenbosch, E. Steinbach, A delay compensation approach for pan-tilt-unit-based stereoscopic 360 degree telepresence systems using

head motion prediction, in: 2018 IEEE International Conference on Robotics and Automation (ICRA), 2018.

[2] Q. Wang, W. Jiao, R. Yu, M.T. Johnson, Y.M. Zhang, Modeling of human welders' operations in virtual reality human–robot interaction, IEEE Robotics and Automation Letters 4 (3) (2019) 2958–2964.

[3] Ming Gao, Ralf Kohlhaas, J. Marius Zöllner, Contextual learning and sharing autonomy to assist mobile robot by trajectory prediction, in: 2016 IEEE International Symposium on Safety, Security, and Rescue Robotics (SSRR), IEEE, 2016, pp. 274–275.

[4] P.K. Artemiadis, K.J. Kyriakopoulos, EMG-based teleoperation of a robot arm using low-dimensional representation, in: IEEE/RSJ International Conference on Intelligent Robots and Systems, 2015.

[5] Jake J. Abbott, Allison M. Okamura, Virtual fixture architectures for telemanipulation, in: IEEE International Conference on Robotics and Automation, vol. 2, IEEE, 2003, pp. 2798–2805.

[6] Chuanjiang Xia, Qingxuan Jia, Gang Chen, Fine manipulation simulation of GAMMA300 7-DOF robot based on virtual fixture, in: IEEE Conference on Industrial Electronics and Applications (ICIEA), IEEE, 2017, pp. 1327–1331.

[7] Fredrik Rydén, Howard Jay Chizeck, Sina Nia Kosari, H.H. King, Blake Hannaford, Using Kinect and a haptic interface for implementation of real-time virtual fixture, 2011.

[8] Yuheng Fan, Chenguang Yang, Xinyu Wu, Improved teleoperation of an industrial robot arm system using leap motion and MYO armband, in: 2019 IEEE International Conference on Robotics and Biomimetics (ROBIO), IEEE, 2019, pp. 1670–1675.

[9] Chunhua Ren, Tianda Fu, Meilin Zhou, Xiaoming Hu, Low-cost 3-d positioning system based on sEMG and MIMU, IEEE Transactions on Instrumentation and Measurement 67 (4) (2018) 876–884.

[10] O.R. Seryasat, F. Honarvar, Abolfazl Rahmani, et al., Multi-fault diagnosis of ball bearing using FFT, wavelet energy entropy mean and root mean square (RMS), in: IEEE International Conference on Systems, Man and Cybernetics, IEEE, 2010, pp. 4295–4299.

[11] D.G. Lloyd, T.S. Buchanan, A model of load sharing between muscles and soft tissues at the human knee during static tasks, Journal of Biomechanical Engineering 118 (3) (1996) 367–376.

CHAPTER SEVEN

Task learning of teleoperation robot systems

7.1. Introduction

In a human-in-the-loop teleoperation system, the human is a major factor. The human can send control commands to the remote robot and also receive feedback from the remote robot and the environment to adjust the control inputs in time. However, because teleoperation systems suffer from the model uncertainty and the lack of system transparency, it is a challenge for the operator to simply control the remote robot. Researchers have proposed many control methods to address the model uncertainty and time delay problems. For example, to simultaneously solve the kinematics and dynamics of teleoperation systems, Yang et al. proposed radial basis neural network controllers and wave variables to compensate for the effects caused by model uncertainties and time delay [1]. In addition, methods such as adaptive fuzzy control and impedance control are effective solutions to the problems of remote operation model uncertainty and time delay [2]. However, with these methods it is difficult to obtain an accurate model of the system in case of an uncertain environment model and uncertain load. At present, robot learning methods have become an effective way to study the relationship between the operators, the robots, and the interactive tasks. During teleoperation, operators are faced with tremendous operational pressure and task load. In some extreme cases, the operator needs to control each joint of the robot to guide it to interact with the environment. Therefore, it becomes very practical to simplify the relationship between the operators, the robots, and the interaction tasks by enabling the robot to learn human operation skills through robot learning methods and to reduce the operator's workload. For these reasons, this chapter is based on the study of how to improve the efficiency of the robot learning and the automatic generation of tasks.

In this chapter, a robot task trajectory learning method based on machine learning is proposed. First, the dynamic time regularization method is used to deal with the problem of time mismatch of the demonstration data. By temporal regularization, the demonstrated data can be normalized in the same time domain. Second, a hybrid Gaussian model is used to program key parameters such as the trajectory of the task to describe the relationship between the robot, the operator, and the interaction task. Finally, a hybrid regression model is used to generate continuous, smooth, and state-dependent skill models based on the demonstrated tasks. Compared with traditional teleoperation, the teleoperated robot can autonomously replicate the learned human skills and complete the tasks based on the generated skill models after several demonstrations. This improves the efficiency to a certain extent and does not require human involvement. At the same time, the robotic learning approach allows the teleoperation system to reduce human intervention and the operator can concentrate on the decision-making work.

In this chapter, we also put forward a robot teaching method which uses a virtual teleoperation system based on visual interaction and uses neural learning based on an extreme learning machine (ELM). More specifically, in our work, Kinect V2, a Microsoft second-generation motion capture device, is used to track the human body motion and the hand state. A learning algorithm based on ELM [3] which can teach the robot to learn some skills from humans is developed. Compared with robot learning based on other neural networks, this ELM requires fewer training samples and has a high generalization capacity [4].

Additionally, we propose a novel human–robot interaction (HRI) perception mechanism and learning algorithm to improve the manipulation performance of teleoperation robots. The robot can improve its intelligence by learning the human operating behavior, and the learned behaviors include not only the trajectory of the movement but also the stiffness of the human during the task. By learning human behaviors, teleoperated robots can independently perform repetitive or uncertain tasks. To effectively characterize the human–robot collaboration (HRC) tasks and improve the intelligence of the robot, a probabilistic statistics-based robot learning framework is proposed. The framework divides the HRC task into two phases: the learning phase and the reproduction phase. In the learning phase, the HRC task can be learned based on the hidden semi-Markov and hybrid Gaussian model approaches to build a generative task model based on motion trajectories. In the reproduction phase, the task trajectory of the generative model can be corrected based on the hidden semi-Markov and hybrid Gaussian regression methods to replicate the results from the learned task.

Inspired by the robot learning algorithms and the muscle activation-based control method, in this chapter, a nonlinear regression model with physiological interface is constructed to reduce the workload of the human operator to improve efficiency of the teleoperation system and capture detail information of the remote operation environment. First, the dynamic time warping (DTW) method is employed to align the demonstration data in the same time scale before proceeding with other steps in the demonstration process. Second, a human-centric interaction method based on muscle activation is developed to collect the operation information and to actively capture the remote environment online. Besides, the locally weighted regression (LWR) method is proposed to model the teleoperated prescribed task based on the collected task trajectories and the human stiffness. Finally, the feasibility and efficiency of the proposed method are verified by experiments using physical robots.

7.2. System description

As shown in Fig. 7.1, the teleoperation system introduced in this section can be divided into three parts: human demonstrations, robot learning, and robot execution.

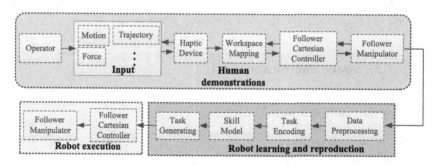

Figure 7.1 Robot learning framework of the teleoperation system.

(1) Human demonstration module

In this module, the human operator controls the follower robot to perform tasks through the position commands from the leader haptic device. During this process, the trajectories including position and velocity of the robot's end-effector are recorded. Since the kinematic structure of the leader device and the follower robot are different, we adopt the following coordinate transformation:

$$
\begin{bmatrix} \vec{x} \\ \vec{y} \\ \vec{z} \end{bmatrix} = \vec{O}_{sm} \times \left\{ \vec{\Lambda} \begin{bmatrix} \vec{x}_m \\ \vec{y}_m \\ \vec{z}_m \end{bmatrix} + \vec{b} \right\}, \tag{7.1}
$$

where $\vec{x}$, $\vec{y}$, $\vec{z}$ and $\vec{x}_m$, $\vec{y}_m$, $\vec{z}_m$ are the position vectors of the follower and the leader in task space, respectively, $\vec{O}_{sm}$ is the transformation matrix, $\vec{\Lambda}$ is the scale factor to adjust the workspace between the leader and the follower, and $\vec{b}$ is the position correction term in the x-, y-, and z-directions.

(2) Robot learning and reproduction module

The robot learning and reproduction module can be divided into three parts: data pre-processing, task coding, and task generating. The main purpose of data pre-processing is to standardize the demonstrated data into a unified time domain. Task coding is performed to decompose the skills of the operator into a form that the computer can understand and generate a learning skill model for the follower device according to the new situation.

(3) Robot execution module

This module enables the robot to follow the trajectory generated by the robot learning module. In other words, the follower robot performs the general tasks learned from the demonstrated robot trajectory.

7.3. Space vector approach

In 3D space, the distance between point $A(x_1, y_1, z_1)$ and point $B(x_2, y_2, z_2)$ can be calculated by the following formula.

$$
d = \sqrt{(x_2 - x_1)^2 + (y_2 - y_1)^2 + (z_2 - z_1)^2}. \tag{7.2}
$$

The vector $\overrightarrow{AB}$ can be expressed as $\overrightarrow{AB} = (x_2 - x_1, y_2 - y_1, z_2 - z_1)$, $d = |\overrightarrow{AB}|$. In 3D space, the cosine law can be used to calculate the angle between two joints. A joint in Kinect coordinates can be expressed as a vector. Joint 1 is $\overrightarrow{AB}$, and joint 2 is $\overrightarrow{BC}$. In 3D space, the angle between two vectors can be calculated by using the cosine law, so the angle between two joints can be expressed as

$$
cos(\overrightarrow{AB}, \overrightarrow{BC}) = \frac{\overrightarrow{AB} \cdot \overrightarrow{BC}}{|\overrightarrow{AB}| \cdot |\overrightarrow{BC}|}. \tag{7.3}
$$

According to the above equation, the coordinates returned by the camera can be converted to corresponding vectors. Then the angles of the joints can be calculated using the cosine law.

We can use the position coordinates obtained by Kinect to construct a geometric model of the human left arm (Fig. 7.2).

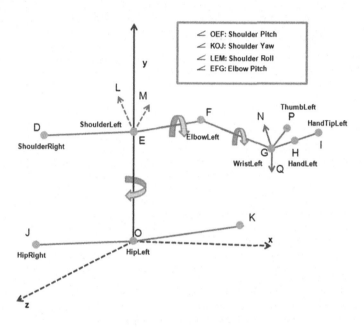

Figure 7.2 A geometry model of the human left arm.

To calculate the shoulder pitch angle $\angle$OEF, we get the vectors $\overrightarrow{OE}$ and $\overrightarrow{EF}$ from three points and then calculate the angle. Using the same method, we can get the elbow pitch $\angle$EFG. Projecting Points D, O, and F to the plane XOZ, we can calculate the shoulder yaw angle $\angle$KOJ.

Based on the above information, we can calculate $\overrightarrow{LE}$ and $\overrightarrow{ME}$ through the following formula:

$$\begin{cases} \overrightarrow{LE} = \overrightarrow{EF} \times \overrightarrow{FG}, \\ \overrightarrow{ME} = \overrightarrow{EF} \times \overrightarrow{DE}, \end{cases} \tag{7.4}$$

so the shoulder roll angle $\angle$LEM can be calculated. In the same way, we can calculate the elbow roll angle, which is the angle between $\overrightarrow{LE}$ and $\overrightarrow{GN}$, and the hand yaw, which is the angle between $\overrightarrow{GN}$ and $\overrightarrow{GQ}$.

7.4. DTW-based demonstration data processing

As a well-known algorithm to process time series signals, the DTW method can not only calculate the similarity between two temporal sequences, but also align their time lines by dynamic programming.

Consider two time series with different length, $x = [x_1, x_2, ..., x_m]$ and $y = [y_1, y_2, ..., y_n]$. Suppose the matrix WP is the match path between two given time series built by DTW. We have

$$WP = \begin{bmatrix} wp_{11} & wp_{21} & ... & wp_{L1} \\ wp_{12} & wp_{22} & ... & wp_{L2} \end{bmatrix}, \tag{7.5}$$

where L is the length of the match path and $wp_{k1} \in \{1, 2, ..., m\}$, $wp_{k2} \in \{1, 2, ..., n\}$, $k \in \{1, 2, ..., L\}$.

Constrained by match path WP, the distance between x and y can be expressed as

$$\rho\left(WP, x, y\right) = \sum_{k=1}^{L}(x_{wp_{k1}} - y_{wp_{k2}})^2, \tag{7.6}$$

where $\left[wp_{k1}, wp_{k2}\right]^T$ is the kth column of the matrix WP. The match path WP is subject to the following conditions:

(1) Boundary condition. We have $\left[wp_{k1}, wp_{k2}\right] = [1, 1]$ and $\left[wp_{L1}, wp_{L2}\right] = [m, n]$.

(2) Monotonicity condition. For $k = 1, 2, ..., L$, $wp_{11} \leq wp_{21} \leq ... \leq wp_{k1}$ and $wp_{12} \leq wp_{22} \leq ... \leq wp_{k2}$.

(3) Step size condition. For $k = 1, 2, ..., L-1$, $wp_{k+1,1} - wp_{k1} \leq 1$ and $wp_{k+1,2} - wp_{k2} \leq 1$.

There are a few paths that meet the above conditions. The DTW method is used to find the optimal one, which is expressed as

$$DTW(x, y) = min(\rho(WP, x, y)). \tag{7.7}$$

Solving $DTW(x, y)$ by dynamic programming, an accumulated cost matrix R can be obtained, with a dimension of $m \times n$. $R(i, j)$ can be expressed as

$$R(i,j) = (x_i - y_i)^2$$

$$+ \begin{cases} 0 & \text{if } i = 1 \text{ and } j = 1, \\ R(i, j-1) & \text{if } i = 1 \text{ and } j > 1, \\ R(i-1, j) & \text{if } i > 1 \text{ and } j = 1, \\ min\{R(i-1,j), R(i,j-1), R(i-1,j-1)\} & \text{otherwise.} \end{cases} \quad (7.8)$$

After processing the sample with the DTW algorithm, we can align them on the time line, as shown in Fig. 7.3.

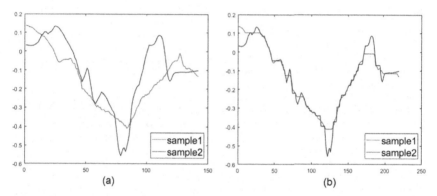

Figure 7.3 (a) The two sample values of the robot joint angle. (b) Two sample values are aligned by DTW.

7.5. Task learning

7.5.1 GMM and GMR

Reinforcement learning (RL) and hidden Markov models (HMMs) are widely used to effectively encode task trajectories [5,6]. Nevertheless, due to the need for good timeliness to encode a task, the RL method is limited because its search space is too large for relatively complex tasks. Besides, the process of task encoding is continuous, while the HMM method must be interpolated based on discrete sets. Therefore, we propose a method based on Gaussian Mixture Model (GMM) and Gaussian Mixture Regression (GMR) to deal with the abovementioned shortcomings in robot learning based on its appropriate search space and continuity.

(1) Task encoded by GMM

After pre-processing, N normalized demonstration sample sequences can be obtained, which can be expressed as

$$X = \{X^i\}_{i=1}^N = \{x_t^i, \dot{x}_t^i\}_{t=0, i=1}^{T,N}, \tag{7.9}$$

where $X \in \mathbb{R}^M$ consists of positions and velocities of the follower end-effector, $x_t \in \mathbb{R}^{M1}$ and $\dot{x}_t \in \mathbb{R}^{M2}$ are the positions and velocities of the follower, respectively, and we have $M = 6$ ($M_1 = 3$ and $M_2 = 3$) dimensions with respect to the observations.

GMM encodes the demonstration task from the observation sequence [7,8], and its probability density function $p(X^i)$ can be expressed as

$$p(X^i) = \sum_{j=1}^K p(j)\mathcal{N}(X^i|\mu^j, \sigma^j), \tag{7.10}$$

where μ^j and σ^j are the mean values and variances for the observations, $p(j)$ is the prior information for the K Gaussian components in the jth step satisfying

$$\sum_{j=1}^K p(j) = \sum_{j=1}^K \pi(j) = 1, \tag{7.11}$$

and $N(X^i|\mu^j, \sigma^j)$ is the Gaussian distribution of the observations given as

$$N(X^i|\mu^j, \sigma^j) = \frac{1}{\sqrt{(2\pi)^3 |\sigma^j|}} exp\{-\frac{(X^i - \mu^j)^T \sigma^{j-1}(X^i - \mu^j)}{2}\}. \tag{7.12}$$

Substituting Eqs. (7.11)–(7.12) into Eq. (7.10), we get

$$\begin{aligned} N(x_t^i, \dot{x}_t^i|\Theta) &= p(X^i) \\ &= \sum_{j=1}^K p(j)\mathcal{N}(X^i|\mu^j, \sigma^j) \\ &= \sum_{j=1}^K \pi(j)N(x_t^i, \dot{x}_t^i|\mu^j, \sigma^j) \\ &= \sum_{j=1}^K \pi(j)\frac{1}{\sqrt{(2\pi)^3 |\sigma^j|}} exp\{-\frac{(X^i - \mu^j)^T \sigma^{j-1}(X^i - \mu^j)}{2}\}, \end{aligned} \tag{7.13}$$

where $\Theta = \{\Theta^j\}_{j=1}^K = \{\pi^1, \mu^1, \sigma^1, ..., \pi^j, \mu^j, \sigma^j, ..., \pi^K, \mu^K, \sigma^K\}$ are the parameters of the Gaussian component. In this chapter, the demonstrated observations are regarded as independent Gaussian distribution. According to the demonstrated observations, the GMM parameters can be estimated by employing the expectation-maximization (EM) method. The values of the EM method are initialized by using the K-means clustering algorithm.

(2) Task generated by GMR

According to the demonstrated observation sequence, the GMR method is employed to generate a generalized task model encoded by GMM [9–11]. By using the joint probability distribution of the observation sequence, the mean matrix and covariance matrix can be described as

$$\mu^j = \begin{bmatrix} \mu_i^j \\ \mu_o^j \end{bmatrix} = \begin{bmatrix} \mu_x^j \\ \mu_{\dot{x}}^j \end{bmatrix}, \tag{7.14}$$

$$\sigma_j = \begin{bmatrix} \sigma_i^j & \sigma_{io}^j \\ \sigma_{oi}^j & \sigma_o^j \end{bmatrix} = \begin{bmatrix} \sigma_{xx}^j & \sigma_{x\dot{x}}^j \\ \sigma_{\dot{x}x}^j & \sigma_{\dot{x}\dot{x}}^j \end{bmatrix}. \tag{7.15}$$

The estimated values of the conditional distribution for output data $\hat{X}_o^j$ can be computed according to the conditional output X_o^j and given X_i as follows:

$$\begin{aligned} \hat{X}_o^j &= \mu_o^j + \sigma_{oi}^j (\sigma_i^j)^{-1} (X_i - \mu_i^j) \\ &= \mu_{\dot{x}}^j + \sigma_{\dot{x}x}^j (\sigma_{xx}^j)^{-1} (X_i - \mu_x^j), \end{aligned} \tag{7.16}$$

$$\begin{aligned} \hat{\sigma}_o^j &= \sigma_o^j - \sigma_{oi}^j (\sigma_i^j)^{-1} \sigma_{io}^j \\ &= \sigma_{\dot{x}}^j - \sigma_{\dot{x}x}^j (\sigma_{xx}^j)^{-1} \sigma_{x\dot{x}}^j, \end{aligned} \tag{7.17}$$

where μ_x^j and $\mu_{\dot{x}}^j$ are the mean of the position and the velocity, respectively, and $\sigma_{\dot{x}x}^j$ is the covariance between velocity and position.

Inspired by Refs. [12,13], the conditional distribution of output data $\hat{X}_o$ for the Kth Gaussian component is

$$\begin{aligned} \hat{X}_o &= \sum_{j=1}^K \beta^j \hat{X}_o^j \\ &= \sum_{j=1}^K \beta^j \{\mu_{\dot{x}}^j + \sigma_{\dot{x}x}^j (\sigma_{xx}^j)^{-1} (X_i - \mu_x^j)\}, \end{aligned} \tag{7.18}$$

$$\beta^j = \frac{p(X_i|j)}{\sum_{j=1}^K p(X_i|j)}$$

$$= \frac{p(j)p(X_i|j)}{\sum_{j=1}^K p(j)p(X_i|j)}, \tag{7.19}$$

$$\hat{\sigma}_o = \sum_{j=1}^K \beta^{j2} \hat{\sigma}_o^j$$

$$= \sum_{j=1}^K \beta^{j2} \{\sigma_{\dot{x}}^j - \sigma_{\dot{x}x}^j (\sigma_{xx}^j)^{-1} \sigma_{x\dot{x}}^j\}. \tag{7.20}$$

Therefore, a generated task model is obtained based on GMR. The generated motion from the learned model can perform smoothly regardless of the inverse kinematics problem of the follower, greatly improving the real-time ability of the telerobotic systems for automated tasks.

With the conditional probability of GMR and a current given position x, the desired velocity can be expressed as

$$\hat{\dot{x}} = \sum_{j=1}^K h^j \{[A^j \quad b^j][x \quad 1]^T\}, \tag{7.21}$$

where

$$A^j = \sigma_{\dot{x}x}^j (\sigma_{xx}^j)^{-1}, \tag{7.22}$$

$$b^j = \hat{\mu}_{\dot{x}}^j - \sigma_{\dot{x}x}^j (\sigma_{xx}^j)^{-1} \mu_x^j, \tag{7.23}$$

$$h^j = \frac{\mathcal{N}(x|\mu_x^j, \sigma_{xx}^j)}{\sum_{j=1}^K \mathcal{N}(x|\mu_x^j, \sigma_{xx}^j)}, \tag{7.24}$$

where $j = 1, ..., K$ in Eqs. (7.22)–(7.24).

According to Euler integration, the desired position in Cartesian space at time t is updated based on the computed desired velocity $\hat{\dot{x}}$:

$$\hat{x} = x|_{t-1} + \sum_{j=1}^K h^j \{[A^j \quad b^j][x \quad 1]^T\}. \tag{7.25}$$

7.5.2 Extreme learning machine

We can generate commands for the robot through an artificial dynamic system. The dynamic system can be expressed by the following equation:

$$\dot{s} = f(s) + \varepsilon, \tag{7.26}$$

where s denotes the robot's end-effector position or joint angles and $\dot{s}$ is the first derivative of s. The dataset is $\{s, \dot{s}\}_{t=0}^{T_1, \cdots, T_L}$, where T_i is used to represent different time points; ε is a zero mean Gaussian noise [14]. The goal is to obtain an estimation of $\hat{f}$ from f.

We use a method based on ELM to achieve this goal, which is more efficient than the traditional learning algorithm under the same conditions. To use ELM in the teleoperation system, the goal is to learn the mapping $f : s \to \dot{s}$ based on the dataset $\{s, \dot{s}\}_{t=0}^{T_1, \cdots, T_L}$. As shown in Fig. 7.4, for a neural network with a hidden layer, the input layer has n nodes, which is the dimension of s. In the hidden layer, the target function is as follows:

$$\dot{s} = o = \sum_{i=1}^{L} \beta_i f_i(s) = \sum_{i=1}^{L} \beta_i g(\omega_i^T s + b_i), \qquad (7.27)$$

where g is an activation function, $W = (\omega_1, \omega_2, ..., \omega_L)^T$ is the input weights with dimension $L \times d$, d is the distance between point $A(x_1, y_1, z_1)$ and point $B(x_2, y_2, z_2)$ as shown in Eq. (7.2), $\beta = (\beta_1, \beta_2, ..., \beta_L)^T$ is the output weights, also with dimension $L \times d$, $b = (b_1, b_2, ..., b_L)$ is the hidden layer biases, and $\omega_i^T s$ is inner product of W and s.

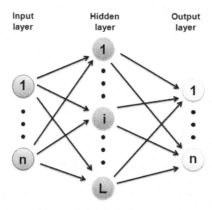

Figure 7.4 Extreme learning machine.

The single hidden layer neural network is used to minimize the output error. ELM solves the problems as follows:

$$\min_{\beta} \|H\beta - O\|, \qquad (7.28)$$

$$H(\omega_1, \omega_2, ..., \omega_L, b_1, b_2, ..., b_L, s_1, s_2, ..., s_L)$$

$$= \begin{bmatrix} g(\omega_1^T s_1 + b_1) & \cdots & g(\omega_L^T s_1 + b_L) \\ \vdots & \ddots & \vdots \\ g(\omega_1^T s_L + b_1) & \cdots & g(\omega_L^T s_L + b_L) \end{bmatrix}, \tag{7.29}$$

where $H(\omega_1, \omega_2, ..., \omega_L, b_1, b_2, ..., b_L, s_1, s_2, ..., s_L)$ is the output of the hidden layer node and $O = (o_1, o_2, ..., o_L)^T$ is the expected output. In the system, O is the target value which is generated by demonstration.

Once the input weights and hidden layer biases are fixed, the output matrix of the hidden layer H can be clearly determined. Then the problem of training the single hidden layer neural network can be translated into a problem of solving a linear system. The solution is as follows:

$$\beta = H^+ O, \tag{7.30}$$

where H^+ is the Moore–Penrose generalized inverse of matrix H.

For any $t \in R$, the activation functions $g(t)$ should be continuous and continuously differentiable. We use activation functions which satisfy

$$\begin{cases} g(t) = 0, & t = 0, \\ g(t) > 0, & \forall t \in R. \end{cases} \tag{7.31}$$

With the purpose of meeting the above conditions, a bipolar sigmoid function is used:

$$g(t) = \frac{2}{1 + e^{-t}} - 1. \tag{7.32}$$

The sigmoid functions are continuous and continuously differentiable, so they do not have an impact on the performance of ELM.

7.5.3 Locally weighted regression (LWR)

Because the demonstration data are high-dimensional, it is difficult to find a function to describe the demonstration data globally. The LWR algorithm is adopted to explore the approximate function locally between the input and the output for the given aligned demonstration data.

The local relationship between the output $y(i)$ and the input $x(i)$ can be obtained as follows:

$$Y = \frac{\sum_{i=1}^{N} w(i) y(i)}{\sum_{i=1}^{N} w(i)}, \tag{7.33}$$

where $y(i)$ denotes the output values of the ith local model, $w(i)$ represents the weights of $y(i)$, and N indicates the number of datasets for the demonstration. Based on the receptive fields of the model, the weight w_i can be computed in Gaussian forms as follows:

$$w(i) = exp\{-\frac{1}{\vartheta}(x - x(i))^T(\boldsymbol{x} - x(i))\}, \qquad (7.34)$$

where x is the fitting position and $\vartheta > 0$ is used to adjust the Gaussian weight. In this study, $\vartheta = 2$. Eq. (7.34) shows that the weight value becomes larger as the input $x(i)$ is near the fitting position $\boldsymbol{x}$.

For the given input data $x(i)$, the log likelihood for the probability $p(y(i)|x(i))$ can be presented as

$$
\begin{aligned}
\ell(\Xi) &= log\{\prod_{i=1}^{N} p(y(i)|x(i))\} \\
&= \sum_{i=1}^{N} log(\frac{w(i)}{2\pi})^{\frac{k}{2}} \\
&\quad - \frac{1}{2}\sum_{i=1}^{N} w(i)\{y(i) - \Xi x(i)\}^T\{y(i) - \Xi x(i)\},
\end{aligned}
\qquad (7.35)
$$

where Ξ is a parameter vector of the LWR method.

According to Refs. [15,16], the maximum of $\ell(\Xi)$ can be obtained by minimizing $H(\Xi)$ as follows:

$$H(\Xi) = \frac{1}{2}\sum_{i=1}^{N} w(i)\{y(i) - \Xi x(i)\}^T\{y(i) - \Xi x(i)\}, \qquad (7.36)$$

where

$$\Xi_{min} = argmin\ \{\frac{1}{2}\sum_{i=1}^{N} w(i)\{y(i) - \Xi x(i)\}^T\{y(i) - \Xi x(i)\}\}, \qquad (7.37)$$

$$
\begin{aligned}
\frac{\partial \ell(\Xi)}{\partial \Xi} &= \sum_{i=1}^{N} w(i)(\Xi x(i) - y(i))x(i)^T \\
&= \Xi\sum_{i=1}^{N} w(i)x(i)x(i)^T - \sum_{i=1}^{N} w(i)y(i)x(i)^T.
\end{aligned}
\qquad (7.38)
$$

To set Eq. (7.38) to be 0, we get

$$W = \begin{bmatrix} w(1) & 0 & \cdots & 0 \\ 0 & w(2) & \cdots & 0 \\ & \cdots & & \\ 0 & 0 & \cdots & w(N) \end{bmatrix}. \tag{7.39}$$

With Eqs. (7.38)–(7.39), the nonlinear model can be obtained as

$$\Xi = B^T WA (A^T WA)^{-1}, \tag{7.40}$$

where

$$A = \begin{bmatrix} x(1)^T \\ x(2)^T \\ \cdots \\ x(N)^T \end{bmatrix}, \tag{7.41}$$

$$B = \begin{bmatrix} y(1)^T \\ y(2)^T \\ \cdots \\ y(N)^T \end{bmatrix}. \tag{7.42}$$

After the task is learned by using the LWR method, the follower robot can be operated according to the learned task trajectories and learned human stiffness.

7.5.4 Hidden semi-Markov model

In this section, we select the end-effector of a follower device as the input device to perform a specific task by training with different initial conditions, e.g., different initial locations. Then, we can obtain a task learning model by using the hidden semi-Markov model (HSMM) method.

The proposed task generative frame is shown in Fig. 7.5. In the *learning* phase, HSMM-GMM is used to obtain the task model parameters. In the *reproduction* phase, HSMM-GMR is used to reproduce the telerobot behavior.

(1) Parameters of the task generative model

The input data come from the collections of positions x and velocities $\dot{x}$ of the end-effector of robots.

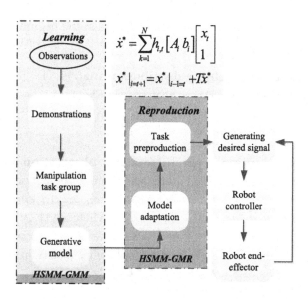

Figure 7.5 Proposed task generative model.

We can regard the position of the follower device $\xi \in \mathbb{R}^D$ as an observation sequence:

$$
\begin{cases}
\xi = \begin{bmatrix} \xi_t^I \\ \xi_t^O \end{bmatrix}, \\
x = x_l,
\end{cases}
\tag{7.43}
$$

where ξ^I is the position of the follower robot manipulator with the input components, ξ^O are output components, and $x_l \in \mathbb{R}^3$ represents the position of the follower.

The task generative model has six elements:

(1) State of the model. Suppose there are N states or N hidden states in the HSMM, i.e., $S = [S_1, S_2, S_3, \cdots, S_N]$. At time t, the hidden state is $q_t \in [S_1, S_2, S_3, \cdots, S_N]$.

(2) Number of observations. M indicates the number of observations. A set V incorporating M observations is denoted as $V = [v_1, v_2, v_3, \cdots, v_M]$. At time t, the observation is $o_t \in [v_1, v_2, v_3, \cdots, v_M]$.

(3) Initial probability distribution vector. $\pi = [\pi_1, \pi_2, \pi_3, \cdots, \pi_N]$ represents the initial probability distribution of the model, which defines the probability distribution of each hidden state at the beginning of calculation $\pi_i = P(q_t = S_i) \geq 0$, $1 \leq i \leq N$, and satisfies $\sum_{i=1}^{N} \pi_i = 1$.

(4) Transition probability distribution matrix. $A = [a_{ij}]_{N \times N}$ is the transition probability distribution matrix for hidden state i at time $(t-1)$ to hidden state j at time t, i.e., $a_{ij} = P(q_{t+1} = S_j \mid q_t = S_i)$, $1 \leq i, j \leq N$. It satisfies $\sum_{j=1}^{N} a_{ij} = 1$, $1 \leq i \leq N$.

(5) Gaussian mixture parameters and observation probability matrix. We use the Gaussian joint probabilities to represent the output probability distribution for observation, i.e., $\mu = (\mu_1, \mu_2, \mu_3, \cdots, \mu_i, \cdots, \mu_N)$ and $\sigma = (\sigma_1, \sigma_2, \sigma_3, \cdots, \sigma_i, \cdots, \sigma_N)$, where $i \in [1, N]$. Therefore, the observation probability at time t for state S_i is $p_i(t) = \mathcal{N}(t; \mu_i, \sigma_i)$. $B = b_i(k)_{N \times M}$ indicates the observation probability matrix in state S_i, and $b_i(k) = P(o_t = v_k, q_t = S_i)$, $1 \leq i \leq N$, $1 \leq k \leq M$.

(6) Probability density function for state dwell time. We train the model c times using a set $c \in \{1, 2, \cdots, C\}$, and we have $\mu^C = (\mu_1^C, \mu_2^C, \mu_3^C, \cdots, \mu_N^C)$ and $\sigma^C = (\sigma_1^C, \sigma_2^C, \sigma_3^C, \cdots, \sigma_N^C)$, where μ^C and σ^C are the mean values and variances of the Gaussian duration probability distribution, respectively. The state duration probability density function $p_i^C(t)$ of state i is $p_i^C(t) = \mathcal{N}(t; \mu_i^C, \sigma_i^C)$.

Therefore, the proposed model with N states is parameterized by

$$\lambda = \{\pi, A, \{\mu, \sigma\}, \{\mu^C, \sigma^C\}\}_{i,j}^N. \tag{7.44}$$

(2) Initialization of the task generative model

With the above analysis, it is necessary to initialize the task model λ, that is, to recalculate the values of the given task model parameters. Generally speaking, the form of the Markov chain is determined by model parameters π and A. But in the presence of Markov chains, the initial values of π and A have almost no effect on the final convergence effect of the model. Among the parameters of the actual model, the processing of the state duration probability density function $p_i^C(t)$ will affect the final result of the model, but the impact is also limited. Therefore, the model $\{\pi, A, p_i^C(t)\}$ can be initialized randomly or with equal values. The other two parameters $\{\mu, \sigma\}$ of the model have a greater impact on the convergence of the model. Generally, we use the K-means clustering method to initialize μ, σ [17,18].

(3) Training method

The purpose of task model training is to calculate the parameters of the model $\lambda = \{\pi, A, \{\mu, \sigma\}, \{\mu^C, \sigma^C\}\}$ through the expected-maximum method according to the existing observation sequence $V = [v_1, v_2, v_3, \cdots, v_t]$ [19]. Assuming that the observation sequence in this chapter obeys an independent Gaussian probability distribution, the training process of the task model is shown in Fig. 7.6.

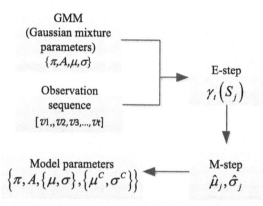

Figure 7.6 Training of the model.

According to the given observation sequence, the probability $\gamma_t(S_j)$ of state S_j at time t is evaluated as follows:

$$\gamma_t(S_j) = P(S(t) = s_j \mid V, \lambda) = \frac{p(V, S(t) = s_j \mid \lambda)}{p(V \mid \lambda)}. \tag{7.45}$$

We can recalculate the parameters of the task model based on $\gamma_t(S_j)$ as follows:

$$\begin{cases} \hat{\mu}_j = \frac{\sum_{t=1}^{T} \gamma_t(S_j) v_t}{\sum_{t=1}^{T} \gamma_t(S_j)}, \\ \hat{\sigma}_j = \frac{\sum_{t=1}^{T} \gamma_t(S_j)(v_t - \hat{\mu}_j)(v - \hat{\mu}_j)^T}{\sum_{t=1}^{T} \gamma_t(S_j)}. \end{cases} \tag{7.46}$$

(4) Observation constituent

We define the position of the end-effector of the follower robot ξ_t as the observation sequence of the model, which can be expressed as follows:

$$\xi_t = \begin{bmatrix} \xi_t^I \\ \xi_t^O \end{bmatrix}, \tag{7.47}$$

$$\mu_i = \begin{bmatrix} \mu_i^I \\ \mu_i^O \end{bmatrix} = \begin{bmatrix} \mu_i^x \\ \mu_i^{\dot{x}} \end{bmatrix}, \tag{7.48}$$

$$\sigma_i = \begin{bmatrix} \sigma_i^I & \sigma_i^{IO} \\ \sigma_i^{OI} & \sigma_i^O \end{bmatrix} = \begin{bmatrix} \sigma_i^{xx} & \sigma_i^{x\dot{x}} \\ \sigma_i^{\dot{x}x} & \sigma_i^{\dot{x}\dot{x}} \end{bmatrix}, \tag{7.49}$$

where ξ_t includes the input variable ξ_t^I and the output variable ξ_t^O and μ_i and σ_i are the mean and variance related to the input and output, respectively.

(5) Probability calculation

With the given model $\lambda = \{\pi, A, \{\mu, \sigma\}, \{\mu^C, \sigma^C\}\}$ and observation sequence $V = [v_1, v_2, v_3, \cdots, v_M]$, the probability of mission model observation sequence V can be calculated by the forward algorithm [20]

$$\alpha_{i,t} = P(v_1, v_2, v_3, \cdots, v_t, i_t = q_i \mid \lambda) \tag{7.50}$$

For $t = 1, 2, 3, \cdots, T - 1$, we obtain

$$\alpha_{i,t} = \sum_{j=1}^{N} \sum_{c=1}^{\min(c^{max}, t-1)} \alpha_t^j a_{ji} p_i^C(c) \prod_{s=(t-c+1)}^{t} \mathcal{N}(\xi_c \mid \hat{\mu}_i, \hat{\sigma}_i), \tag{7.51}$$

where $i = 1, 2, 3, \cdots, N$, $\alpha_{i,t}$ is the forward probability $P(V \mid \lambda)$ in state i at time t with $V = (v_1, v_2, v_3, \cdots, v_t)$, and a_{ji} is the transition probability from state j at time t to state i at time t.

(6) Reproduction

According to the probability calculation of HSMM-GMR, a normalized parameter h_i can be defined to describe the influence of state i, whose expression is as follows:

$$h_{i,t} = \frac{\alpha_{i,t}}{\sum_{k=1}^{N} \alpha_{k,t}}, \tag{7.52}$$

$$\alpha_{k,t} = (\sum_{k=1}^{N} \alpha_{k,t-1} a_{ki}) \mathcal{N}(\xi_t \mid \hat{\mu}_i, \hat{\sigma}_i), \tag{7.53}$$

where $\alpha_{i,t}$ is defined in Eq. (7.51).

From Eq. (7.52) and Eq. (7.53), we have

$$h_{i,t} = \frac{\alpha_{i,t}}{\sum_{k=1}^{N} \alpha_{k,t}} = \frac{\mathcal{N}(\xi_t \mid \hat{\mu}_i, \hat{\sigma}_i)}{\sum_{k=1}^{N} \mathcal{N}(\xi_t \mid \hat{\mu}_k, \hat{\sigma}_k)}. \tag{7.54}$$

Referring to the work in Refs. [21,22], the current target velocity $\dot{x}$ can be obtained as

$$\dot{x} = \sum_{i=1}^{N} h_{i,t}\{[A_i \quad b_i][x_t \quad 1]^T\}, \tag{7.55}$$

with

$$A_i = \sigma_i^{xx}(\sigma_i^{xx})^{-1}, \tag{7.56}$$

$$b_i = \hat{\mu}_i^x - \sigma_i^{xx}(\sigma_i^{xx})^{-1}\mu_i^x. \tag{7.57}$$

According to Eqs. (7.55)–(7.57), we can compute the velocity of the end-effector. Assuming that the position of the end-effector of the follower robot at time t is known, we can obtain the position at time $t + 1$ by the following formula:

$$
\begin{aligned}
x|_{i=t+1} &= x|_{i-1=t} + \mathcal{T}\dot{x} \\
&= x|_{i-1=t} + \sum_{i=1}^{N} h_{i,t}\{[A_i \quad b_i][x_t \quad 1]^T\},
\end{aligned} \tag{7.58}
$$

where $\mathcal{T}$ is the length of a single iteration time step.

7.6. Experimental case study

7.6.1 Cleaning experiment

(1) Experimental setup

In this experiment, the teleoperation robot system in Fig. 7.7, which consists of the Touch X device and the Baxter robot, is used to verify the proposed task learning method. In the cleaning task, a cardboard attached to the end of Baxter's right arm is used as a cleaning tool and a cube in the garbage bucket is used as rubbish. A human operator controls the right arm of the Baxter robot via the Touch X device to sweep the rubbish into the garbage bucket.

(2) Pre-processing demonstrated data

We conducted the cleaning task 10 times with different initial places. Fig. 7.8 shows the demonstrated trajectories and the aligned trajectories by using the DTW method. Figs. 7.8(a–c) and 7.8(d–f) show the follower's

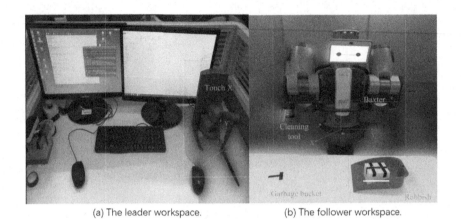

(a) The leader workspace.   (b) The follower workspace.

Figure 7.7 Experimental setup for a cleaning task.

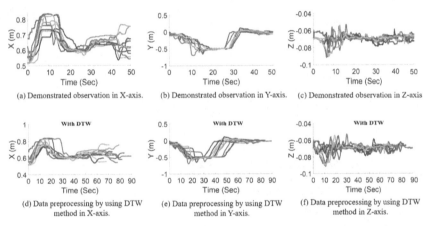

(a) Demonstrated observation in X-axis.   (b) Demonstrated observation in Y-axis.   (c) Demonstrated observation in Z-axis

(d) Data preprocessing by using DTW   (e) Data preprocessing by using DTW   (f) Data preprocessing by using DTW
     method in X-axis.   method in Y-axis.   method in Z-axis.

Figure 7.8 (a) Collection of observations without DTW in the x-direction from 10 demonstrations. (b) Collection of observations without DTW in the y-direction from 10 demonstrations. (c) Collection of observations without DTW in the z-direction from 10 demonstrations. (d) Collection of observations with DTW in the x-direction. (e) Collection of observations with DTW in the y-direction. (f) Collection of observations with DTW in the z-direction.

end–effector trajectory before and after applying the DTW method, respectively. It can be concluded that in each demonstration a different amount of time is needed to complete the cleaning task. The results show that the DTW method improves the smoothness of the demonstrated trajectories and ensures the synchronicity of the motion in the same time domain.

(3) Task learning and execution

In this experiment, the GMM method is employed to encode the cleaning task. Fig. 7.9(a–c) shows that the related GMM model can be encoded by five states. In the generation phase, the desired position can be obtained according to the GMR method. The results of the generation task are plotted in Fig. 7.9(d–f) based on the learned model. There is a smooth generated trajectory in the x-, y-, and z-directions. It is noted that the generated trajectories could be adjusted according to the feedback controller of the follower from different initial positions, as shown in Fig. 7.9(d–f).

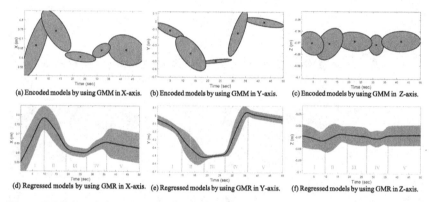

(a) Encoded models by using GMM in X-axis.   (b) Encoded models by using GMM in Y-axis.   (c) Encoded models by using GMM in Z-axis.

(d) Regressed models by using GMR in X-axis.   (e) Regressed models by using GMR in Y-axis.   (f) Regressed models by using GMR in Z-axis.

Figure 7.9 (a) Task encoding using GMM in the x-direction. (b) Task encoding using GMM in the y-direction. (c) Task encoding using GMM in the z-direction. (d) Generation task using GMR in the x-direction. (e) Generation task using GMR in the y-direction. (f) Generation task using GMR in the z-direction.

The robot execution process of the cleaning task is presented in Fig. 7.10. In Fig. 7.10(a), the robot is controlled from an initial position which is different from that in the training steps. In order to describe the execution process of the follower, we manually segment the process into five phases (phases I–V). In phase I, the motion process corresponds with the first motion in the generation period. Accordingly, Fig. 7.10(b–f) indicates the cleaning task process (four phases) corresponding to the another four motions, which are generated using GMR. In the robot execution process, the cleaning tasks are successfully performed using the GMM-GMR method.

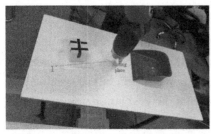

(a) Robot execution process from the starting place to phase I for a cleaning task.

(b) Robot execution process from phase I to phase II for a cleaning task.

(c) Robot execution process from phase II to phase III for a cleaning task.

(d) Robot execution process in phase III pausing for a moment for a cleaning task.

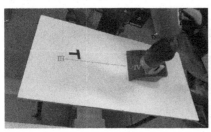

(e) Robot execution process from phase III to phase IV for a cleaning task.

(f) Robot execution process from phase IV to phase V for a cleaning task.

Figure 7.10 The robot execution process for a cleaning task from phase I to phase V.

7.6.2 Kinect-based teleoperation experiment

(1) Collection and processing of experimental data

In this part, we design a simulation scene to verify the effectiveness of learning from demonstration based on ELM. After acquiring the position of the operator from the Kinect device, the Baxter robot in V–REP simulation environment moves its arm and pulls down a square workpiece on the desk, transferring the workpiece to a conveyor belt (Fig. 7.11). We have recorded the robot joint angle, which changes with time. Finally, we have repeated this process 25 times.

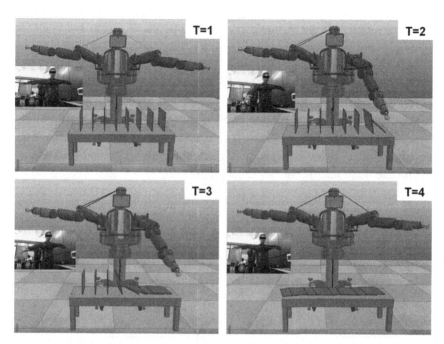

Figure 7.11 Controlled by the human with a Kinect device, the Baxter robot can push down the workpiece on the desk.

As the time caused of the robot movement is different in each experiment, the length of the experimental data is variable. To ensure that the length of each dataset is the same, some data are interpolated. We obtain a sample with dimension 142×3 from each experiment. To confirm the interpolation method has no effect on the effect of exercise, we use each dataset to control the Baxter robot. As expected, the robot can reproduce the learned trajectory.

(2) Data training and learning based on ELM

We implement the algorithm based on ELM in MATLAB®. To find the optimal number of hidden layer nodes, we perform a test in which the number of nodes changes from 1 to 100. The result shows that the situation with 12 nodes is optimal. Then we train the input data with ELM, getting three groups of output data. As shown in Fig. 7.13, they are the values of the robot's corresponding joint angles. In the following experiment, the data are sent to V–REP through MATLAB. Controlled by MATLAB, the Baxter robot in V–REP can reproduce the trajectory which is provided by the ELM. In another scene we design, the Baxter robot can transfer a workpiece from a conveyor belt to another one, as shown in Fig. 7.14. It performs well in repetitive tasks.

Additionally, from Figs. 7.12 and 7.13, we can see that these robot trajectories are mainly related to S0 and S1. During the downward movement of the robot arm, the minimum value of S0 is approximately equal to −0.4 radians, and the maximum value of S1 is approximately equal to 0.6 radians. Therefore, the robot can learn to move a trajectory by teleoperation based on human–robot interaction. The neural network based on ELM can obtain the main features of this trajectory.

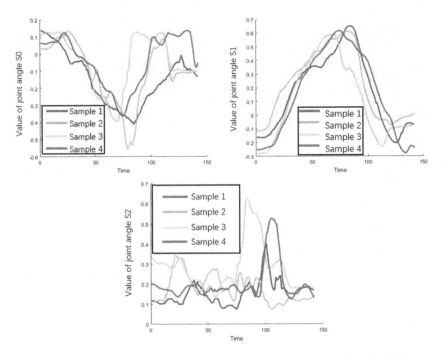

Figure 7.12 Four randomly selected training samples. It can be seen that the difference between datasets is very large.

7.6.3 Cleaning experiment with LWR

(1) Experimental framework

Fig. 7.15 exhibits the experimental process of the proposed method. In the demonstration and learning processes, the human operator directly controls the follower. The proposed approach is trained by collecting prescribed task trajectories and human stiffness. In the robot execution process, the

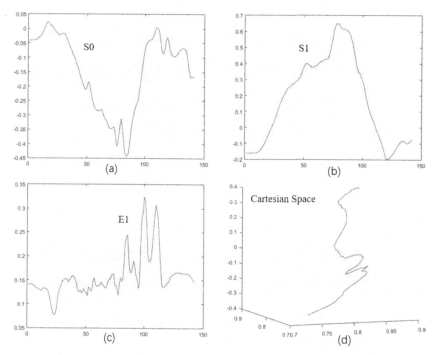

Figure 7.13 The output data of the ELM. (a–c) Joint angles of S0, S1, and E1. (d) Trajectory of the Baxter arm moving downward in Cartesian space. All angles are given in radians.

follower executes the task based on the learned trajectories and learned human stiffness.

(2) Experimental results

A learned task model was calculated by using the LWR algorithm including the demonstrated task trajectories and the human muscle stiffness. The learned Cartesian trajectories are shown in Figs. 7.16(a)–7.16(c). The learned trajectories are bold for the convenience of inspection. Fig. 7.16(d) illustrates the learned human muscle stiffness in the process of demonstrations. Figs. 7.16(a)–7.16(d) demonstrate that the model of demonstrated trajectories and human muscle stiffness can be built by the proposed LWR method.

Based on the learned model using the LWR algorithm, the Baxter robot can execute the sweep task automatically with the learned human muscle stiffness and the learned trajectories. The sweep task fully indicates the

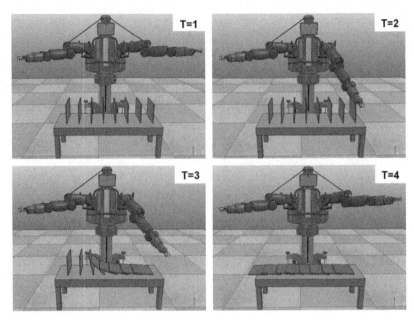

Figure 7.14 Through learning and training by ELM, the Baxter robot can push down the workpiece on a conveyor belt autonomously.

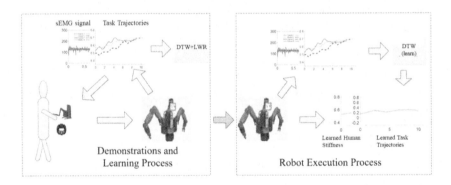

Figure 7.15 Experimental process of the proposed method.

dexterity of human arms in the process of operation. In this experiment, human muscle stiffness was employed based on muscle activation to reflect the interaction information in the demonstration phase.

Fig. 7.17 shows the process of robot execution for the sweep task. By employing the proposed LWR algorithm with physiological interface, the sweep task is performed successfully within 10 s.

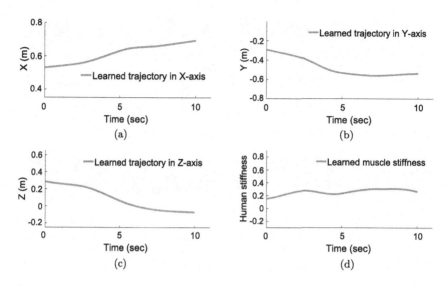

Figure 7.16 Learning using the LWR method.

(3) Discussion

First of all, the proposed LWR method is used to construct a nonlinear model for a specific task. Compared to traditional robot learning methods [23,24], the proposed method combines the task trajectories with muscle stiffness during the demonstration process.

Moreover, the human operator can regulate the muscle stiffness according to the remote environments in the human–robot interaction. That is to say, the variability of muscle stiffness represents the operational characteristics of the human operator. Therefore, a model to describe the relationship between the prescribed task, task trajectories, and muscle stiffness is very useful in the learning process. Thus, the learned model can be used to improve the efficiency of the reproduction of the repetitive task.

Noteworthily, the experimental results in this work are preliminary and the experiments are conducted mainly to show the feasibility and effectiveness of the proposed method. The stiffness of the human operator varies from person to person. Moreover, even for the same person, the stiffness property at different times could be different (time-varying). Therefore, it is hard to find a baseline to compare the results under "with stiffness" and "without stiffness" conditions. Different people may have different stiffness values and thus give different motion models. Therefore, it is difficult to define fair comparison criteria for different/multiple experimental sub-

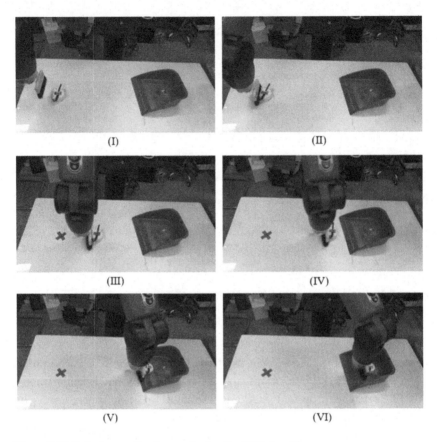

Figure 7.17 Robot execution by employing the LWR algorithm.

jects. No problem is foreseen in the generalization of the proposed method to different subjects as the implementation procedures do not change and the steps are not affected by the specific human subject. However, the obtained specific muscle activation/stiffness pattern(s) will surely be different from person to person.

7.6.4 Drawing experiment

The purpose of this experiment is to evaluate the performance of the presented method in a simple task.

(1) Experimental setting

In this experiment, a typical drawing task is performed by the teleoperation system as presented in Fig. 7.18. In this experiment, a pen is attached to the endpoint of the follower right arm as a drawing tool. A human operator manipulates the leader device to teleoperate the end-effector of the follower by the leader device to perform a drawing task. We adopt $C = 3$ and $N = 18$ of the HSMM in the *learning* phase. We then perform a drawing task in a 210 mm × 297 mm (A4) space.

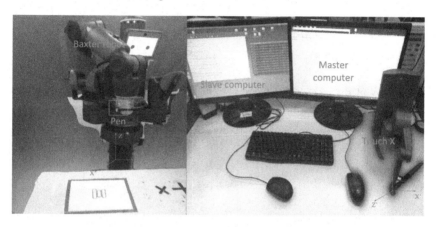

Figure 7.18 Experimental setup for a drawing task.

(2) Results and analysis

The motion trajectories and stiffness profile of the drawing task are shown in Figs. 7.19(a)–7.19(c). The thin curves indicate the demonstration results. The bold curve represents the reproduction results. The *reproduction* phase can be divided into six steps (I–VI). In steps I and II, the robot begins to perform a subtask (drawing task 1). In steps II–III, the end-effector of the follower leaves the paper for another drawing operation. Similarly, drawing task 2 and drawing task 3 are performed in steps III–IV and steps V–VI, respectively. During the *learning* phase, humans' stiffness is variable, which follows from the drawing operation. As shown in Fig. 7.19(d), humans' stiffness maintains a high level in steps I–II, III–IV, and V–VI.

In Fig. 7.20, the telerobot performs the drawing task by using a reproduced stiffness. From Fig. 7.20(a–f), It can be concluded that the drawing tasks are successfully performed by employing the proposed method.

7.7. Conclusion

This chapter presents several frameworks for task learning of teleoperation robots to explore the relationship between teleoperation robots,

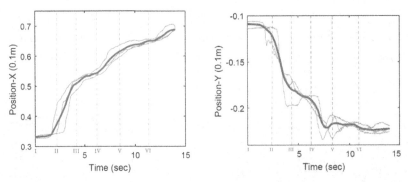

(a) Position of the robot in the x-direction during the drawing task.   (b) Position of the robot in the y-direction during the drawing task.

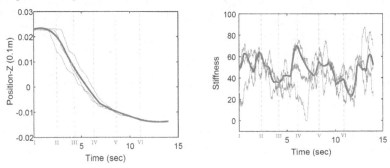

(c) Position of the robot in the z-direction during the drawing task.   (d) Stiffness of the human during the drawing task.

Figure 7.19 Reproduced trajectory and stiffness during the drawing task. Panel (a) shows the reproduced x-position. Panel (b) shows the reproduced y-position. Panel (c) shows the reproduced z-position. Panel (d) shows the reproduced stiffness during the drawing task.

operators, and tasks. First, we propose a novel task learning framework that utilizes a teleoperation approach and GMM to encode the trajectory of the teleoperation robot end-effector. In this framework, GMR generates task models based on different schematic starting points. Based on this, a novel haptic myoelectric perception mechanism and a robot learning framework based on a HSMM coupled with Gaussian hybrid theory are proposed. Integration of the DTW method and the LWR method enabled the robots to learn the task trajectories and human muscle stiffness simultaneously after human demonstrations using a teleoperation system. The remote robot can execute the task according to the learned task trajectories and learned stiffness. Additionally, we have developed a virtual teleoperation system based

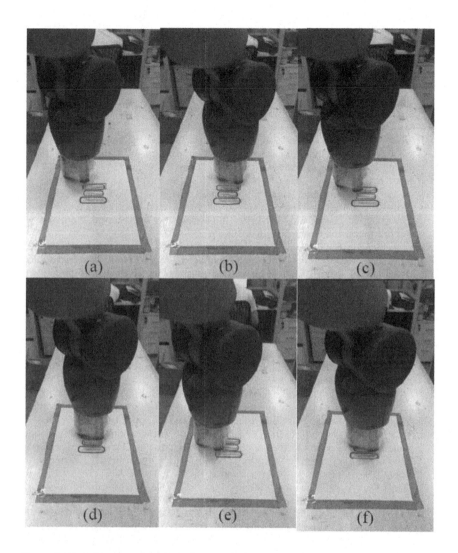

Figure 7.20 Robot execution for a drawing task.

on visual interaction. Human body motion is used to control the robot's arms, and gestures are used to control the beginning and end of the simulation. Finally, the effectiveness of the proposed methods was demonstrated by experiments.

References

[1] Xingjian Wang, Chenguang Yang, Junpei Zhong, Rongxin Cui, Min Wang, Teleoperation control for bimanual robots based on RBFNN and wave variable, in: 2017 9th

International Conference on Modelling, Identification and Control (ICMIC), IEEE, 2017, pp. 1–6.

[2] Wei He, Yiting Dong, Adaptive fuzzy neural network control for a constrained robot using impedance learning, IEEE Transactions on Neural Networks and Learning Systems 29 (4) (2017) 1174–1186.

[3] Chenguang Yang, Kunxia Huang, Hong Cheng, Yanan Li, Chun-Yi Su, Haptic identification by elm-controlled uncertain manipulator, IEEE Transactions on Systems, Man, and Cybernetics: Systems 47 (8) (2017) 2398–2409.

[4] Luka Peternel, Tadej Petrič, Jan Babič, Human-in-the-loop approach for teaching robot assembly tasks using impedance control interface, in: IEEE International Conference on Robotics and Automation, IEEE, 2015, pp. 1497–1502.

[5] Jens Kober, J. Andrew Bagnell, Jan Peters, Reinforcement learning in robotics: A survey, The International Journal of Robotics Research 32 (11) (2013) 1238–1274.

[6] Leonel Rozo, Pablo Jiménez, Carme Torras, Force-based robot learning of pouring skills using parametric hidden Markov models, in: 9th International Workshop on Robot Motion and Control, IEEE, 2013, pp. 227–232.

[7] Leonel Rozo, Danilo Bruno, Sylvain Calinon, Darwin G. Caldwell, Learning optimal controllers in human–robot cooperative transportation tasks with position and force constraints, in: 2015 IEEE/RSJ International Conference on Intelligent Robots and Systems (IROS), IEEE, 2015, pp. 1024–1030.

[8] Matteo Saveriano, Dongheui Lee, Learning motion and impedance behaviors from human demonstrations, in: 2014 11th International Conference on Ubiquitous Robots and Ambient Intelligence (URAI), IEEE, 2014, pp. 368–373.

[9] Firas Abi-Farraj, Takayuki Osa, Nicoló Pedemonte, Jan Peters, Gerhard Neumann, Paolo Robuffo Giordano, A learning-based shared control architecture for interactive task execution, in: 2017 IEEE International Conference on Robotics and Automation (ICRA), IEEE, 2017, pp. 329–335.

[10] Sylvain Calinon, Paul Evrard, Elena Gribovskaya, Aude Billard, Abderrahmane Kheddar, Learning collaborative manipulation tasks by demonstration using a haptic interface, in: 2009 International Conference on Advanced Robotics, IEEE, 2009, pp. 1–6.

[11] Maura Power, Hedyeh Rafii-Tari, Christos Bergeles, Valentina Vitiello, Guang-Zhong Yang, A cooperative control framework for haptic guidance of bimanual surgical tasks based on learning from demonstration, in: 2015 IEEE International Conference on Robotics and Automation (ICRA), IEEE, 2015, pp. 5330–5337.

[12] Mattia Racca, Joni Pajarinen, Alberto Montebelli, Ville Kyrki, Learning in-contact control strategies from demonstration, in: 2016 IEEE/RSJ International Conference on Intelligent Robots and Systems (IROS), IEEE, 2016, pp. 688–695.

[13] Sylvain Calinon, Danilo Bruno, Milad S. Malekzadeh, Thrishantha Nanayakkara, Darwin G. Caldwell, Human–robot skills transfer interfaces for a flexible surgical robot, Computer Methods and Programs in Biomedicine 116 (2) (2014) 81–96.

[14] S. Mohammad Khansari-Zadeh, Aude Billard, Learning stable nonlinear dynamical systems with Gaussian mixture models, IEEE Transactions on Robotics 27 (5) (2011) 943–957.

[15] Yuandong Sun, Huihuan Qian, Yangsheng Xu, Robot learns Chinese calligraphy from demonstrations, in: 2014 IEEE/RSJ International Conference on Intelligent Robots and Systems (IROS 2014), IEEE, 2014, pp. 4408–4413.

[16] Jin Wang, Liang Chih Yu, K. Robert Lai, Xuejie Zhang, Locally weighted linear regression for cross-lingual valence-arousal prediction of affective words, Neurocomputing 194 (C) (2016) 271–278.

[17] Leonel Rozo, João Silvério, Sylvain Calinon, Darwin G. Caldwell, Learning controllers for reactive and proactive behaviors in human–robot collaboration, Frontiers in Robotics and AI 3 (30) (2016) 1–11.

[18] Ajay Kumar Tanwani, Sylvain Calinon, Small variance asymptotics for non-parametric online robot learning, 2016.

[19] Ajay Kumar Tanwani, Sylvain Calinon, A generative model for intention recognition and manipulation assistance in teleoperation, in: 2017 IEEE/RSJ International Conference on Intelligent Robots and Systems (IROS), IEEE, 2017, pp. 43–50.

[20] Ajay Kumar Tanwani, Sylvain Calinon, Learning robot manipulation tasks with task-parameterized semitied hidden semi-Markov model, IEEE Robotics and Automation Letters 1 (1) (2016) 235–242.

[21] Sylvain Calinon, Antonio Pistillo, Darwin G. Caldwell, Encoding the time and space constraints of a task in explicit-duration hidden Markov model, in: 2011 IEEE/RSJ International Conference on Intelligent Robots and Systems (IROS), IEEE, 2011, pp. 3413–3418.

[22] Sylvain Calinon, Aude Billard, Statistical learning by imitation of competing constraints in joint space and task space, Advanced Robotics 23 (15) (2009) 2059–2076.

[23] Stefan Schaal, Learning from demonstration, in: Advances in Neural Information Processing Systems, 1997, pp. 1040–1046.

[24] Hsien-Chung Lin, Te Tang, Yongxiang Fan, Yu Zhao, Masayoshi Tomizuka, Wenjie Chen, Robot learning from human demonstration with remote lead through teaching, in: 2016 European Control Conference (ECC), IEEE, 2016, pp. 388–394.

Index

Printed in the United States
by Baker & Taylor Publisher Services